MARINE
ALGAE
of the west coast
of Florida

MARINE
ALGAE
of the west coast
of Florida

CLINTON J. DAWES

University of Miami Press

Coral Gables, Florida

Clinton J. Dawes, Ph. D.
Professor of Botany and Assistant Chairman, Department of Biology, University of South Florida, Tampa, Florida

Copyright © 1974 by
University of Miami Press

Designed by Anne Hertz

Manufactured in the United States of America

Illustration credits:
photographs:
Dr. Harry E. Calvert—figs. 7, 8, 11, 14, 19, 23, 24.
Mr. Donald P. Cheney—figs. 4, 6, 9, 10, 12, 13, 16-18, 20-22.
Dr. Clinton J. Dawes—figs. 1, 2, 5, 15, 25.
Mr. Wayne R. Fagerberg—figs. 3, 81, 82.
line drawings:
Miss Linda Baumhardt—figs. 26-52, 54-57, 59-64, 66, 67, 73-78, 80
Mr. Frank Hoff—figs. 53, 58, 65, 68-72, 79

Library of Congress Cataloging in Publication Data
Dawes, Clinton J
    Marine algae of the west coast of Florida.

    Bibliography: p.
    1. Marine algae—Florida. 2. Marine algae—
Mexico, Gulf of. I. Title.
QK571.5.F6D38   589'.39'2634   73-22107
ISBN 0-87024-258-X

Cover drawing of Caulerpa mexicana *by Linda Baumhardt*

To my students,
without whose help this book could not have been written,
and to Dr. Harold J. Humm,
who introduced me to the marine plants of Florida

Contents

Contents

Illustrations

Tables

Preface

After publishing *Marine Algae in the Vicinity of Tampa Bay* in 1967, I realized the need for a complete, detailed, and yet simple, guide to the marine algae of the entire west coast of Florida. This book is designed to fill that need, with chapters on the algal representations of the divisions Cyanophyta, Rhodophyta, Chlorophyta, and Phaeophyta. Other chapters include discussions of marine algae, marine angiosperms, collection sites, habitats, and plant communities. The underlying theme of the book, marine plant ecology, stresses the importance of plants and their adaptations to the marine environment.

The text is organized and written for the interested layman, as well as for high school and college students. The species keys for each division, whenever possible, rely upon clear-cut vegetative characters. The species descriptions contain a minimum of taxonomic detail, terminology, and literature and a maximum of habitat and ecological information.

Drawings and photographs illustrate the common species from the four algal divisions. My thanks go to Miss Linda Baumhardt and Mr. Frank Hoff, who executed the habit sketches, and to Mr. Donald Cheney and Mr. Harry Calvert, who contributed many of the photographs of blue-green algae.

Numerous friends and fellow scientists have aided me in preparing this book, but special thanks should go to Dr. Harold J. Humm (Director, Marine Science Department, University of South Florida, St. Petersburg, Florida); Dr. Sylvia A. Earle (Associate Scientist, Los Angeles County Museum, Los Angeles, California), and Dr. Arthur Mathieson (Director, Jackson Estuarine Marine Laboratories, University of New Hampshire, Durham, New Hampshire).

I would also like to express gratitude to the many students who took my courses in marine botany and phycology and who have aided me in writing the text. Students who directly contributed to this book are Mr. David Ballantine, Mr. Don Cheney, Mr. Wayne Fagerberg, Mr. Nat Eiseman, Mr. Andrew Rehm. A special acknowledgment goes to Mrs. Pat Carey and Mrs. Dianne Geschwender for their constructive help in typing the manuscript.

C.J.D.
Tampa, Florida

MARINE
ALGAE
of the west coast
of Florida

1

Physical characteristics
of marine habitats

This book provides students and others concerned with marine science with a descriptive and ecological account of the marine flora along the west coast of Florida to approximately 70 miles offshore Tampa Bay. For the specific area covered, see p. 4. As an important record of the present marine environment and associated flora, this book will allow comparison with circumstances several years hence. Radical changes in the coastline of Florida are currently being effected through dredge and fill operations and other influences of a rapidly expanding human population. Development of the coastline makes it urgent to investigate existing natural and relatively undisturbed habitats. Sykes (1964) pointed out the problems of dredge and fill operations with relation to destruction of marine habitats and the shrimping industry of Tampa Bay.

HISTORICAL SYNOPSIS

Humm (1973 *b*) has outlined early studies on marine algae of the Caribbean. Taxonomic and ecological information on marine plants and their communities along the west coast of Florida dates from about 1940. Davis (1943), in one of his many publications, dealt thoroughly with the mangrove and salt grass marshes of the west coast. Humm (1956) and Phillips (1962) studied the sea grasses of the Gulf coast. Phillips (1960*a*, 1960*b*) and Phillips and Springer (1960) studied the marine algae of Tampa Bay and nearby offshore reefs. Dawes, Earle, and Croley (1967) and Dawes and van Breedveld (1969) established the presence of a diverse, deepwater tropical flora along the west coast of Florida from Tampa Bay to the Florida Keys. Humm and Taylor (1961), S. E. Taylor (1966), and Earle (1969) also emphasized the existence of a relatively stable offshore tropical flora in the eastern Gulf of Mexico. They showed that the Tampa Bay area marks northern and southern distributional limits of certain seasonal species.

Additional information, including distributional records of plants in Tampa Bay, is given by W. R. Taylor (1936, 1954a and 1954b) and Humm (1963, 1964). Good illustrations and descriptions of most of the species are presented by W. R. Taylor (1960) and Børgesen (1913-1920).

Since 1964 my students and I have made numerous collections, observations, and environmental data determinations along the west coast of Florida. We collected specimens by wading and free diving along the shore and by using skiffs and larger vessels offshore. Deepwater sampling was accomplished by dredging and by diving with scuba gear. Collections made by the Florida Department of Conservation, Marine Laboratories (Bayboro Harbor, St. Petersburg) have been studied, and the herbaria at Florida State University, the University of Florida, and Duke University were also examined for pertinent records. Herbarium preparations, including voucher specimens of new records, have been deposited in the algal herbarium of the University of South Florida, and many duplicates were sent to the U.S. National Museum.

The keys, species descriptions, and many habitat notes for the Chlorophyta were modified from Humm and S. E. Taylor (1961) and for the Phaeophyta from S. E. Taylor (1966). W. R. Taylor (1928, 1960) was a major source for many of the identifications and data included in the species descriptions. Drouet (1968) and Drouet and Dailey (1956), as well as Tilden (1910), Desikachary (1959), and Humm (unpublished) were used as sources for the blue-green algae.

MARINE HABITATS

The West Coast, General. The west coast of Florida in this text is the region between Cedar Key and Cape Romano (latitude 29°10′N to 25°50′N; see p. 3). Tampa Bay lies in the central portion of this area (see p. 4) near the line of division between subtropical (to the south) and warm temperate (to the north) inshore waters (Earle, 1969). The Gulf coast consists of a broken string of barrier islands and mangrove-edged lagoons with occasional marshes behind these offshore islands. The elongate coastal lagoons, which are separated from the Gulf of Mexico and formed by the barrier islands, are con-

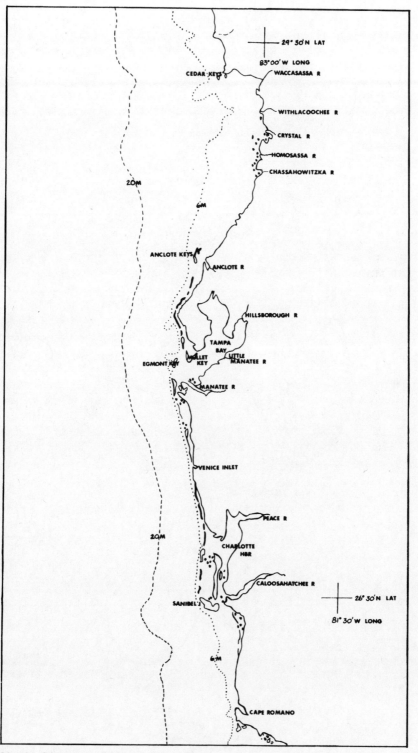

Map 1. Continental shelf around Tampa Bay.

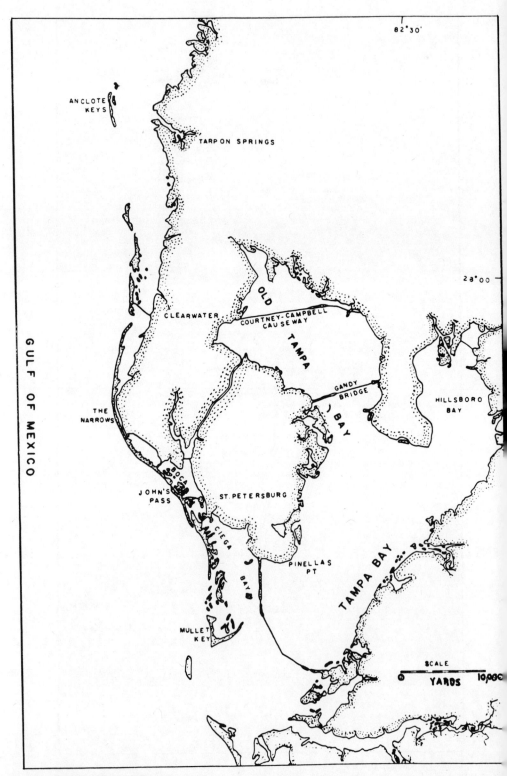

Map. 2. Tampa Bay area.

nected to the Gulf by breaks or passes in the island chain. Boca Ciega Bay, one such lagoon, extends between the mainland and a line of offshore islands in the northwestern part of Tampa Bay, and Sarasota Bay does the same in the southwestern portion of Tampa Bay. Along the open coast the continental shelf is broad, gradually sloping to the west, with an average depth of 30 feet at 5 to 10 miles offshore (see map, p. 3).

The outer coastline, much of Tampa Bay, and the lagoonal coastline consist of sand beaches that offer only limited littoral collection. In general, jetties (Mullet Key, Pass-a-Grille, Ana Maria Island, Venice Inlet), forts (Egmont Key and Mullet Key), and seawalls (Indian Bluff Island, Tierra Verde) support a diversified algal flora showing distinct zonation into supralittoral, littoral, and sublittoral regions. Natural coastal rock apparently is limited to one beach rock formation at Point of Rocks (Siesta Key, Fig. 1). In such an area, a well-developed flora can be found without the typical stunting of algae characteristic of sandy regions. Pilings and rock foundations of bridges (as well as oyster bars) are the only other firm substrata that support a littoral flora.

The west coast of Florida has a number of rivers which originate from large springs a few miles inland and which either flow directly into the Gulf (Waccasassa River, Withlacoochee River, Homosassa River, Chassahowitzka

Fig. 1. A coquina rock outcropping at Point of Rocks (Siesta Key), Florida. A common seaside shrub, the sea grape, is seen in the foreground.

Fig. 2. A view of the mangrove islands and sea grass beds at low tide along the south end of the Skyway at the mouth of Tampa Bay.

River, Weeki Wachee River), into lagoons created by offshore barrier islands (Anclote River, Manatee River, Caloosahatchee River, Cocohatchee River), or into larger bays (Tampa Bay: Hillsborough River, Alafia River, Little Manatee River; Charlotte Harbor: Peace River). At all sites the rivers create estuarine conditions near their mouths.

Littoral marine angiosperms, such as mangroves (Fig. 2), dominate the estuaries south of Tampa Bay, whereas salt marshes (Fig. 3) are common in the estuaries north of Tampa Bay. Coastal marine angiosperm communities offer another type of littoral substrate for the attachment of epiphytic algae (see also chapter 9). Such communities contain many filamentous and parenchymatous forms of algae that show, in some cases, zonal distribution on the prop roots and pneumatophores of mangroves or on the stems of salt marsh plants.

The offshore sea grass beds, so characteristic of the sublittoral zones of warm temperate to tropical waters, are found along the west coast of Florida in depths of 1 to about 40 meters. The sea grass beds are especially well developed in regions where limestone outcroppings are common with a thin overlay of sand and the water transparency is high. The best developed sea grass beds are in waters between Anclote Key and Cedar Key.

The richest algal communities are found in the vicinity of limestone outcroppings. Previous studies by Phillips and Springer (1960), Dawes *et al.*

Fig. 3. A salt grass marsh along a tidal channel at low tide at the Weeki Wachee River. *Spartina alterniflora* forms a frontal band along the channel, and *Juncus roemerianus* comprises the major portion of the salt marsh. A pine-palm hammock is visible in the background.

(1967), Earle (1969), and Dawes and van Breedveld (1969), have determined that the best developed algal flora along the west coast of Florida is sublittoral. The collector will find a flora approaching that of the Florida Keys in abundance, variety, and beauty.

Tampa Bay. Tampa Bay, the central region of this study, presents a wide variety of ecological habitats. It is a large, roughly "Y"-shaped estuary with Old Tampa Bay representing the western fork, Hillsborough Bay the eastern part, and Tampa Bay proper the base or main portion (see map, p. 4). There are many small keys and islands, including spoil banks, in the bay. At the mouth of the bay several islands extend north to New Port Richey (Anclote Key) and south to Fort Myers (Sanibel Island).

The average depth of the bay and lagoons is less than 5 meters, although the numerous channels and passes, both natural and artificially maintained, may reach 30 meters. Within Tampa Bay and in the smaller bays between the mainland and barrier islands, there are extensive littoral and shallow-water oyster bars, sand flats, mud flats, and grass beds. The mean tidal amplitude in the bay varies from 0.4 meter at the lower end of the bay to 0.7 meter at Tampa. Tidal exchange is responsible for the major portion of water circulation (Dragovich and Kelly, Jr., 1964; Dragovich and May, 1962). Several rivers, notably the Hillsborough, Alafia, and Little Manatee, and numerous freshwater streams flow into the bay, creating conditions favorable for the development of the mangroves and salt marshes that border much of the shoreline (Phillips, 1960 *a*). Mangroves line the coast and form islands in the eastern and southern regions of Tampa Bay. Salt marshes are most common in upper Tampa Bay and Boca Ciega Bay. Both of these littoral marine angiosperm communities contain many forms of epiphytic algae.

The sublittoral zone within Tampa Bay supports an extensive algal flora and grass beds (Phillips, 1962). The grass beds are well represented in Sarasota Bay and throughout the shallower regions of Tampa Bay, especially the eastern and southern coasts.

PHYSICAL ECOLOGY

The west coast of Florida and especially Tampa Bay has a humid, subtropical climate. The climate becomes more temperate north of Tampa Bay. The average annual rainfall is 125 cm, and the mean daily temperatures range from 17.6°C to 21.1°C, with water temperature falling closely within this range (Dragovich and Kelly, Jr., 1964). The maximum monthly mean water temperature is about 32°C, and this is usually reached in July or August. The minimum monthly mean water temperature is about 10°C during December or January (U.S. Coast and Geodetic Survey 31-1, 1960). The greatest temperature span over a 12-month period is found in the littoral zone where algae are exposed to direct air temperatures at low tide. The offshore water exhibits a much smaller variation in temperature.

Along the west coast, salinities range from a few parts per thousand in estuaries to that of normal seawater in the open Gulf (34 to 36 parts per thousand). The seasonal salinity changes in Tampa Bay and coastal lagoons are influenced by local rainfall and land runoff. The lowest salinities occur during the late summer to early autumn months, the period of high rain. During periods of heavy rainfall, areas at river mouths display freshwater characteristics (Dragovich, Finucane, and May, 1961; Dragovich and May, 1962).

In Tampa Bay and in estuaries formed by rivers the sediments are a mixture of terrigenous quartz sand and biogenic carbonate, which form a homogenous mass due to shifting by tidal currents (Goodell and Gorsline, 1961). In much of the shallow-water regions the substrate is stabilized by sea grass beds, by mangroves, and by salt marsh plants along the coast. The outer coastline in the vicinity of Tampa Bay consists of elongate barrier islands that are mainly sand. Offshore, north of Tampa Bay, limestone outcroppings become common and supply a continuous firm base for algal attachment. To the south of Tampa Bay the barrier islands are continuous, and the substrate is mainly sand with occasional limestone outcroppings. Wave action along the west coast has been described as moderate with an average coastal wave height of 50 centimeters (Tanner, 1960).

The prop roots of the red mangrove, *Rhizophora mangle;* the pneumatophores of the black mangrove, *Avicennia germinans;* and various salt marsh plants (the black rush, *Juncus roemerianus;* salt grass, *Spartina alterniflora*) provide a stable substrate in otherwise sandy beaches or mud flats. Algal epiphytes are common on the rhizomes of the green alga *Caulerpa* and on the leaves of five species of sea grasses: (1) turtle grass (*Thalassia testudinum*); (2) shoalgrass (*Diplanthera wrightii*); (3) manatee grass (*Syringodium filiforme*); (4) widgeon grass (*Ruppia maritima*); and (5) *Halophila engelmanni*. Previous studies have indicated the importance of the grasses as substrate for marine algae (Humm, 1964). Chapter 8 offers a brief discussion of these mangroves and sea grasses and a description of their vegetative characters.

2

Characteristics of
marine algae

WHAT IS AN ALGA?

Algae are separated into distinct divisions on the basis of their physiological
and morphological differences. In contrast, the angiosperms (flowering
plants), such as sea grasses and mangroves, are placed in one division.

Algae are true plants in that they all have photosynthetic pigments, especially the basic pigment, chlorophyll a. Furthermore, their cells are encased
with a cell wall, and most forms, excluding the blue-green algae, have distinctive plastids and cell vacuoles. Algae are usually differentiated from higher
plants by the lack of true roots, stems, and leaves and the absence of conduct-
ing tissues, such as xylem and phloem. Usually algae have sex organs that are
one-celled, or at least all the cells comprising an algal sex organ function in
reproduction.

A point should be made regarding the "primitiveness" of algae. Algae have
been in existence over one billion years, with some forms almost identical to
those existing today. If we consider that the oceans are relatively stable
systems, the algae appear to be well suited in morphology and reproduction
for such an environment. Some of these organisms, therefore, may be considered equal to the "higher" land plants with regard to adaptation, specializa-
tion, and development.

This book is concerned with four major groups of macroscopic algae that
make up most of the marine benthic flora anywhere in the world: the Cyanophyta (blue-green algae), the Chlorophyta (green algae), the Rhodophyta (red
algae), and the Phaeophyta (brown algae). One other group, the Chrysophyta
(golden-brown algae), has one macroscopic representative, *Vaucheria*. For

convenience, this genus is treated in the key to Chlorophyta and described with the Chlorophyta.

All four groups or divisions, as they are properly called, are large units of algae held together by such common features as pigmentation, types of reserve foods, and reproductive processes. Each division (equivalent to an animal phylum) may have one or more classes, and each class can be subdivided into orders. Each order in turn can be subdivided into families, with each family containing one or more genera and each genus one or more species. For example:

Division: Chlorophyta (green algae)   (phyta)
 Class: Chlorophyceae   (phyceae)
  Order: Siphonales   (ales)
   Family: Caulerpaceae   (aceae)
    Genus: *Caulerpa*
     Species: *prolifera*

The name of a plant is written properly when both genus and species titles appear and when the authority or authorities appear after the name, for example, *Caulerpa prolifera* (Forsskål) Lamouroux. In this case Lamouroux transferred the species *prolifera,* which had been named by Forsskål, from another genus to the genus *Caulerpa.* Note that the endings of division, class, order, and family names are characteristic; consequently, one can easily determine the category to which a name belongs.

This chapter introduces briefly each of the four divisions with regard to pigmentation, reserve food, types of morphology, and cytological and ecological characteristics. Detailed discussions on algal morphology and cytology can be obtained from Prescott (1968) and Dawson (1966). Discussions on various aspects of algal physiology can be found in Lewin (1962).

CYANOPHYTA

The blue-green algae are usually small plants most commonly found in littoral zones. Many of the limestone blocks used to form jetties in the Tampa Bay region are blackened by growths of blue-green algae. The cells also form slippery coatings on pilings and boat ramps, as well as on roots of mangroves and stems of salt marsh plants. The importance of the blue-green algae to the marine environment is difficult to assess. Certainly, they are common organisms in the littoral zone in tropical regions. Blue-green algae form a large portion of the planktonic blooms known as "red tides," and they form coatings on plants and substrates in regions of stagnant water or quiet lagoons. Almost every collection of marine algae will yield some species of blue-green algae, although these algae will not dominate the area. In short, blue-green algae are ubiquitous.

Cytologically, these plants are prokaryotes and therefore must be considered to be bacteria. A prokaryotic cell lacks membraned organelles, such as nuclei, plastids, and mitochondria. Because of their small cell size, from 10 to 70 micrometers (μm) (see Table 1 for measurement explanation), blue-green algae must be studied with compound microscopes. The plant morphology varies from small colonies of cells (see Figs. 6-9) to rather elaborate filaments (see Figs. 15 and 16) that may branch (see Figs. 24 and 25). Filamentous forms may remain clumped together in a macroscopically visible colony.

Table 1. **Units of Measure** (Système International d'Unités)

Unit name	Symbol	Fraction of meter	Number of feet (U.S.)
Meter	m	10^{-0}	3.28
Decimeter	dm	10^{-1}	3.28×10^{-1}
Centimeter	cm	10^{-2}	3.28×10^{-2}
Millimeter	mm	10^{-3}	3.28×10^{-3}
Micrometer	μm	10^{-6}	3.28×10^{-6}
Nanometer	nm	10^{-9}	3.28×10^{-9}

As stated previously, all the algae possess the basic photosynthetic pigment chlorophyll *a*, which reflects green light. In addition, the blue-green algae have two water-soluble pigments called phycobilins, c-phycoerythrin and c-phycocyanin. The blue-reflecting c-phycocyanin pigment is the most common; hence the blue (c-phycocyanin) and green (chlorophyll *a*) coloration. The investigator should be warned that many blue-green algae appear black, brown, or red. The best features for identification of blue-green algae are their cytological characteristics (see key to divisions, chapter 3, first section).

The cells of blue-green algae contain a peripheral pigmented cytoplasm called chromoplasm (no organized plastids) and a rather central, clear nucleoplasm. The cell wall is usually thin, and unless it has a sheath, the wall is indistinct. Many trichomes (rows of cells) have sheaths (see Figs. 12 and 13). The sheaths are easily observed at sites where the trichome terminates and the sheath extends beyond.

Occasionally the cells contain refractive globules, which are the reserve food, cyanophycean starch (Fig. 16). Some genera are characterized by the presence of specialized cells (see Figs. 16 and 20), such as "heterocysts" (large, empty, thick-walled spheres) and "akinetes" (large or elongated thick-walled cells with dense cytoplasm, usually spores). The presence of the following morphological features are used in identification: colonies, trichomes, sheaths, akinetes, heterocysts, and true or false branching. Because of the recent publications of Drouet and Dailey (1956) and Drouet (1968), the taxonomy of the coccoid and unbranched filamentous blue-green algae has been greatly revised and simplified. The studies by Drouet have resulted in a combination of many species and the reduction of genera by 79%. Although many phycologists do not agree with Drouet's classification, my experience in identifying blue-green algae supports this "lumping." The key and species descriptions presented in this book are those proposed by Drouet whenever applicable. In each case, however, the synonyms are also given so that the reader may refer to other major taxonomic treatises on blue-green algae (Tilden, 1910; Desikachary, 1959).

CHLOROPHYTA

There are over 5,000 species of green algae; only about 20% are marine. Four orders of macroscopic green algae are well represented in the tropics and subtropics: the Ulvales, the Siphonocladiales (to include the Cladophorales, *sensu* Dawes, 1969; Chapman, 1954), the Siphonales, and the Dasycladales (Egerod, 1952). Microscopic forms of green algae, especially unicellular forms of the order Volvocales, will not be covered in this text. One colonial member of the Tetrasporales (*Pseudotetraspora*) is included because of its size and frequency of occurrence. Although found primarily in shallow water (0.5 to 15 meters), green algae have been reported in depths of 100 to 150 meters (Dawes and van Breedveld, 1969).

Because the green algae contain the same pigments as higher plants (chlorophylls *a* and *b*, carotenes, xanthophylls), they are grass green in color. The reserve food is starch as found in higher plants; consequently, the iodine test is positive (see discussion on algal identification in chapter 3).

Cytologically, this group has typical cell organelles, including nuclei, mitochondria, and chloroplasts. The pyrenoid, a characteristic feature of many green algae, is found in the chloroplast and is usually surrounded by starch platelets that are formed by it. The cell wall has two major layers, an outer pectic layer and an inner structural layer of such fibrillar substances as cellulose, mannose, or xylose. Morphologically, the marine green algae covered in this book range from small epiphytes to large blades (*Ulva*, Fig. 27) to filaments with cells visible without magnification (*Chaetomorpha*, Fig. 39). Other forms are massive plant bodies consisting of interwoven filaments without cell septations (*Udotea,* Fig. 38). The grass-green color, the presence of starch, and the typical cell organelles will aid in identification.

PHAEOPHYTA

There are approximately 1,500 species of brown algae, and the majority of these species are found in cold temperate to arctic marine waters. Some cold temperate species, such as the giant kelp (*Macrocystis*) of California, are harvested in great amounts for the extract alginate. On the west coast of Florida the brown algal flora is best developed during the winter months when the shallower waters are cool (18°C). Some tropical genera of brown algae are present throughout the year, such as *Sargassum* (Figs. 51, 52), *Dictyota* (Fig. 44), and *Padina* (Fig. 45).

Cytologically the brown algae resemble the green algae; both types have distinct cell organelles and well-defined cell walls. The brown algal cells are uninucleate. The reserve food is not easily observed because it occurs either in a dissolved state as a carbohydrate (laminarin) or as an alcohol (mannitol).

The dominant photosynthetic pigment is chlorophyll *a* and a secondary chlorophyll, chlorophyll *c*. These chlorophylls, however, are masked by a number of xanthophylls (fucoxanthin, neoxanthin, and others) that give the plants a brownish coloration. Since the mixture may result in a green brown color, some confusion can result as to whether the plant is a brown alga. A test to answer such a question is described in the discussion on methods of alga identification in chapter 3.

Morphologically the brown algae demonstrate some of the most complex plant bodies; many of them have stems (stipes), leaves (lamina), floats (bladders), and roots (haptera). There is a high degree of tissue differentiation in the central conduction region (medulla), inner food storage cells (cortex), and outer photosynthetic region (outer cortex and epidermis). An examination of cross sections of stipes or blades of *Sargassum* will demonstrate the tissue complexity of the larger brown algae. The structure of many of the brown algae is much simpler, the plant consisting of filaments or blades.

RHODOPHYTA

Red algae are almost exclusively marine, with less than 2% of the 4,000 species occurring in freshwater. In contrast to the brown algae, the red algae are most abundant in the tropical and subtropical regions of the oceans. A number of species of red algae are harvested because of their extracts (*Gelidium*—agar; *Chondrus*—carrageen). Species of a local genus, *Eucheuma*, (Fig. 63) are harvested in the Philippines for the phycocolloid carrageen. Some of the red algae are harvested as a direct food source, especially in the Far East.

Cytologically the red algae have typical cell organelles, such as plastids, nuclei, and mitochondria. The cell wall may be transparent and appear separated from the actual protoplast. The transparent nature of the cell wall is caused by the solubility of some of the constituents of the cell wall. Pit connections, specialized cytoplasmic or wall thickenings between daughter cells, are characteristic features of advanced red algal cells. The reserve food of the red algae is a nonsoluble carbohydrate called "Floridean starch," which is not associated with plastids, as in the green algae. Morphologically the red algae vary from unicells, uniseriate, unbranched or branched filaments, to large, elegant, feathery plants.

The pigments include the basic photosynthetic pigment chlorophyll *a* and a supporting chlorophyll, chlorophyll *d,* as well as the water-soluble phycobilins. Because the phycobilins differ slightly in chemical structure from those found in the blue-green algae, they are prefixed with r. Whereas the blue-green algae have a dominance of c-phycocyanin (blue-reflecting), the red algae have a dominance of r-phycoerythrin (red-reflecting), hence the red to pink or violet coloration.

3

Keys to the algae

The species descriptions in the following chapters (chapters 4, 5, 6, and 7) include both vegetative and reproductive features. The artificial keys in this chapter, however, are based upon external and internal vegetative structure as far as possible. Reproductive structures are included in the keys only when all else fails. Selection of vegetative characters is necessary since many plants are only found in the vegetative state. Even when plants are found in other states, the reproductive structures of these plants are complex and difficult to determine.

The successful use of the keys will depend mainly upon adequate specimens. Entire plants must be available for examination, not just portions or fragments of plants cast ashore or immature plants that were collected attached to rocks or other objects. Because of the specialized terminology, a glossary is included.

For identifying algae, one should use a compound light microscope and a dissecting microscope. Because of the need for accurate measurements (see Table 1 for measurements), both microscopes should be equipped with calibrated ocular micrometers. The investigator will also need a millimeter ruler, dissecting needles, a sharp razor blade, and 5% HCl (hydrochloric acid) for dissolving calcified cell walls. Cross sections of live or preserved algal specimens can be made by carefully using the index finger of the left hand as a guide while slicing (mincing) the specimen with a fresh, sharp razor blade.

To determine the division to which an alga may belong, one should first examine a section or portion of the specimen under the microscope. The blue-green algae are easily distinguished by the lack of cell organelles and by pigmented cytoplasm (chromoplasm) rather than plastids. The green algae are easily distinguished from the remaining two divisions by their grass-green color. The brown algae are usually olive green or brown, and the red algae are usually rose red to deep purplish red. The red and brown algae overlap in color characteristics since littoral red algae, e.g., *Gracilaria*, may be brown in color and similar brown algae may be green in color. If there is doubt, an alga of the division Rhodophyta will become green when placed in boiling water for a minute. Brown algae treated in the same manner will not become green. If, however, a brown alga is placed in heated 70% alcohol, it will become green; this change will not occur with red algae. In a few instances some of the filamentous brown algae will appear yellow green, especially after fixation in formaldehyde. Consequently, fresh material is the best to use, and, if the color is difficult to determine, this test based on pigment solubility can be used. Another way to differentiate between a green alga and a brown alga that appears green is the starch test. A solution of potassium iodide ($I_2 KI$) added to a green alga will indicate the presence of starch by staining it blue (grams [g] iodide: 2 g KI + 1g I + 25 cc water).

GENERAL KEY TO THE DIVISIONS OF ALGAE

All the keys are dichotomous, that is, each leg consists of two opposing alternatives. The investigator should read both alternatives carefully, make his choice, and move to the leg designated next. Terminology is defined in the glossary.

1. Cells without distinct plastids; pigments blue green, gray green, red green, but not grass green; iodine test for starch negative and motile (flagellated) cells absent. . . Cyanophyta p. 18
1. Cells with plastids; pigments localized in these organelles and not distributed throughout the cell, iodine test negative or positive 2
   2. Cells with grass-green chloroplasts, iodine test for starch positive . . . . Chlorophyta p. 21
   2. Cells having plastids but not grass green in pigmentation, plastids some other color, iodine test for starch negative. 3
3. Plastids olive to dark brown, starch test negative, cells range from cuboidal to lax filaments but all have a distinct cell wall; boiling-water test results in no major color change . . . . .Phaeophyta p. 29
3. Plastids red, violet, gray green to bluish green, cells range in morphology from cuboidal to lax filaments but all have a rather diffuse hyaline cell wall; boiling-water test results in a green coloration . . . . Rhodophyta p. 34

CYANOPHYTA, SPECIES KEY

1. Plants distinctly filamentous trichomes with or without sheath. 11
1. Plants not distinctly filamentous, or if so, penetrating substrate. 2
   2. Plants consisting of solitary cells or colonies, but not forming filaments or showing distinct basal attachment to substrate; dividing equally to produce two same-size daughter cells. 4
   2. Plants originally unicellular, but eventually growing into filamentous mats that penetrate the substratum; dividing unequally to produce a small daughter cell. 3
3. Plants epiphytic or endophytic, consisting of cells 3-7 μm in diameter (or more) with distinct sheaths . . . . .
                                    Entophysalis conferta p. 52
3. Plants lithophytic or growing on nonliving material, composed of a layer of columnar cells 7 or more μm in diameter. . . . . *Entophysalis deusta* p. 52
   4. Cells ovoid to cylindrical before division, dividing in a plane perpendicular to the long axis; plant a colony of cells . . . . *Coccochloris elabens* p. 50
   4. Cells otherwise in shape before division, not dividing in planes perpendicular to the long axis; plant a colony of cells. 5
5. Cells before division in a linear series within the gelatinous matrix, cell division in a single plane through the cell. . . . .*Johannesbaptista pellucida* p. 51
5. Cells otherwise in shape, not forming a linear colony . 6
   6. Cells spherical before division, grouped in gelatinous colonies in regular or irregular patterns in three planes, cell division occurring in all three planes. .7
   6. Cells spherical or otherwise in shape before division but distributed in a flat (sheet) or curved (sphere) matrix, cell division proceeding in two planes perpendicular to each other. 9
7. Plants consisting of cells 6 μm in diameter embedded in a gelatinous matrix that becomes pigmented in aerial habitats . . . . *Anacystis montana* p. 50
7. Plants 6 μm or more in diameter, having a hyaline matrix. 8

8. Cells 8-50 μm in diameter, usually angular . *Anacystis dimidiata* p. 49
8. Cells 6-12 μm in diameter, spherical after division. .
Anacystis aeruginosa p. 49
9. Plant a flat to curved sheet of cells, cells arranged in regular series of rows perpendicular to one another . 10
9. Plant spherical to ovoid, cells radially arranged, pyriform to heart shaped . . . . *Gomphosphaeria aponina* p. 51
10. Cells 1-3.5 μm in diameter, plants small (1-256 celled) . . . . *Agmenellum quadruplicatum* p. 48
10. Cells 4-10 μm in diameter, plants larger . *Agmenellum thermale* p. 48
11. Plants having unbranched trichomes in which no heterocysts or akinetes are found 12
11. Plants having branched or unbranched trichomes in which heterocysts and/or akinetes are found 22
12. Trichomes unicellular or appearing so, may be tightly coiled . . . . *Spirulina subsalsa* p. 58
12. Trichomes multicellular, cells separated by cross walls 13
13. Cytoplasmic granules consistently found on either side of each cross wall 20
13. Cytoplasmic granules not arranged along side walls . . 14
14. Outer wall of terminal cell of trichome thickened. .
16
14. Outer wall of terminal cell of trichome not thickened 15
15. Trichomes attenuated for several to many cells at tips
Porphyrosiphon Notarisii p. 56
15. Trichomes not attentuated at tips except for terminal cell 18
16. Cells one-third as long as broad or longer, plants typically planktonic, protoplasm having numerous pseudovacuoles, the terminal cell bulbous and/or spherical . . . . *Oscillatoria erythraea* p. 54
16. Cells usually less than one-third as long as broad, plants never forming blooms, the terminal cell not bulbous 17
17. Ends of trichome attenuated. . . . . *Oscillatoria princeps* p. 55
17. Ends of trichomes slightly tapered to blunt . . . . *Oscillatoria lutea* p. 55
18. Terminal cell conical, trichomes constricted at cross walls. *Schizothrix arenaria* p. 56
18. Terminal cell spherical to hemispherical, no constriction at cross walls. 19

19. Trichomes up to 3.5 μm in diameter. . . . . . *Schizothrix calcicola* p. 57
19. Trichomes 4-60 μm in diameter. . . . . . . *Schizothrix mexicana* p. 58
  20. Outer wall of terminal cell not thickened, trichome
     usually coiled. . . . . *Arthrospira brevis* p. 52
  20. Outer wall of terminal cell thickened, trichome
     straight or bent to slightly coiled. 21
21. Dense protoplasm and granules along side and cross
   walls of each cell. . . . . *Microcoleus lyngbyaceus* p. 52
21. Dense protoplasm and granules along cross walls only,
   trichomes conspicuously tapered at tips. . . *Microcoleus vaginatus* p. 54
  22. Trichomes branched (having true or false branching,
     or both), usually having both intercalary and basal
     heterocysts; trichomes tapering to a hair. 25
  22. Trichomes unbranched (lacking either true or false
     branches), usually having intercalary heterocysts,
     occasionally these becoming terminal; trichomes
     not tapering to a hair 23
23. Trichomes with a firm sheath; one or more trichomes
   present in the sheath. 24
23. Trichomes without a visible sheath, heterocyst rarely
   terminal, usually intercalary; trichomes usually single,
   not organized into a colony. . . . . *Anabaena fertilissima* p. 59
  24. Trichomes single in a sheath, cells short and discoid
                        Nodularia spumigena var. *major* p. 60
  24. Trichomes more than one in a sheath, at least in
     part of the plant, cells barrel shaped. . . . .
                        Hormothamnion enteromorphoides p. 59
25. Trichomes usually tapering to a hair or slender filament,
   at least at the apex, with basal or intercalary heterocysts, showing at most only false branching. 26
25. Trichomes never tapering to a hair, plants having true
   branching, in some cases (*Mastigocoleus*) side branches
   may become terminal hairs. 33
  26. Trichomes found in colonies that have a spherical to
     hemispherical matrix 27
  26. Trichomes gregarious but having distinct sheaths
     and the colonies not coated with a colonial matrix
                        29
27. Colonial matrix becoming hollow, trichomes 2.5-15 μm
   in diameter at base. 28
27. Colonies solid, trichomes 2.5-5 μm in diameter. . . . *Rivularia atra* p. 62
  28. Trichomes usually more than 8 μm in diameter. . . .
                        Rivularia polyotis p. 63

28. Trichomes usually less than 6 μm in diameter. . . .
. . . . *Rivularia nitida* p. 63
29. Filament consisting of trichomes single, in a sheath, with or without branches; if branches present, they are free. 30
29. Filament consisting of more than one trichome in a sheath, false branches coalescent. . . . . *Dichothrix fucicola* p. 62
    30. Heterocysts basal only, trichomes with obvious tapering from bulbous basal heterocysts, trichomes 12-20 μm in diameter. . . . . *Calothrix confervicola* p. 60
    30. Heterocysts basal and intercalary. 31
31. Trichomes 6-10 to 12 μm in diameter, rarely an intercalary heterocyst, cells isodiametric in middle region of trichome. *Calothrix parietina* p. 61
31. Trichomes 10-40 μm in diameter, cells one-half to one-fourth as long as broad. 32
    32. Trichomes 12-40 μm in diameter, false branching common, sheaths distinct, tapering of trichomes at tip obvious . . . . *Calothrix crustacea* p. 60
    32. Trichomes 10-40 μm in diameter, bases interwoven, tapering of trichome not obvious. . . . . *Calothrix pilosa* p. 61
33. Penetrating shells, limestone, uniseriate trichomes having true branching with terminal heterocysts, some branches hairlike or terminating in a hair . . *Mastigocoleus testarum* p. 63
33. Growing on substrate, main axis becoming biseriate or multiseriate, true branches uniseriate, heterocysts intercalary. . . . . *Stigonema aerugineum* p. 63

CHLOROPHYTA, SPECIES KEY

1. Plants with an erect and prominent main axis, usually in the form of a stalk, which differs from the branches or upper parts; plants often calcified. 2
1. Main axis a cluster of cells, creeping or erect; if erect, the main axis not differing notably from the branches and not calcified. 21
    2. Main axis and branches composed of cactuslike segments; calcified. 3
    2. Main axis and branches not composed of cactuslike segments; plants calcified or not. 6

3. Main axis and site of attachment rapidly becoming obscured and the segments of axis not different from
branches. 4
3. Main axis a distinct stalk that differs with regard to
segment morphology from the segments of upper
branches. 5
   4. Branches originate at random, segments heavily calcified, usually sharply trilobed, original holdfast obscured. . . . . *Halimeda opuntia* p. 82
   4. Branches in one plane, largest segments to 2-3 cm
in diameter, lightly calcified and relatively smooth . .
                        Halimeda discoidea p. 81
5. Segments flattened to swollen and bearing distinct
ridges. . . . . *Halimeda incrassata* p. 81
5. Segments cylindrical. . . . . *Halimeda monile* p. 82
   6. Base of plant (stalk or rhizoids) penetrating sand or
mud. 7
   6. Base attached to stone or shell. 18
7. Stalk bearing one to many flat, fan-shaped blades or a
thickened spongy blade. 8
7. Stalk bearing a tuft of loosely united filaments or filaments organized into distinct branches. 15
   8. Blade soft, spongy; not calcified. 9
   8. Blade or blades flat, firm or leathery; somewhat calcified. 12
9. Blade velvety, hairy, and thick; interior filaments moniliform, 50-70 μm in diameter. . . . . *Avrainvillea nigricans* p. 79
9. Blade thin, smooth, relatively firm; filaments monili-
form or smooth; to 30 μm in diameter. 10
   10. Blade filaments uniform in diameter throughout,
20-30 μm, at best slightly moniliform. . . . *Avrainvillea elliotii* p. 79
   10. Blade filaments more slender at surface than in
center. 11
11. Blades deeply cordate, to 15 cm broad; surface filaments 20-30 μm in diameter and consistently monili-
form. *Avrainvillea asarifolia* p. 78
11. Blades slightly cordate to 7 cm broad, filaments to 30
μm in diameter, consistently smooth, and tapered to-
ward the surface (10-20 μm). . . . . *Avrainvillea levis* p. 79
   12. Plant consisting of numerous small blades arranged
radially around a stalk . . . . *Rhipocephalus phoenix* p. 84

12. Plant consisting of a single, large, but sometimes
proliferous blade. 13
13. Plant having a funnel-shaped, papery blade. . . *Udotea cyathiformis* p. 86
13. Plant with a flat or proliferous blade 14
    14. Blade smooth, the filaments comprising it having
    lateral branches that are terminally flattened. *Udotea flabellum* p. 86
    14. Blade composed of obvious filaments without spe-
    cial lateral projections and thus appearing hairy. . .
                               Udotea conglutinata p. 84
15. Plants consisting of distinct branches, branching dichotomous, each branch having a cortication of utricles and
internal slightly moniliform filaments. . *Pseudocodium floridanum* p. 84
15. Plants having no distinct branches, filaments finally free
and without cortication 16
    16. Filaments 100-200 μm in diameter. . . . . *Penicillus capitatus* p. 83
    16. Filaments larger than 200 μm in diameter. . . . 17
17. Filaments 400-500 μm in diameter. . . . . . *Penicillus lamourouxii* p. 84
17. Filaments 800 or more μm in diameter. . . . *Penicillus dumetosus* p. 84
    18. Stalk with a flattened or concave disk on top. . .19
    18. Most of stalk or all of it clothed with whorled, mono-
    siphonous, dichotomously branched ramuli . *Batophora oerstedi* p. 94
19. Segments of disk fewer than 30, lightly calcified; ends
of segments obtuse or truncate . . . . *Acetabularia farlowii* p. 93
19. Disk segments more than 30, lightly calcified; ends of
segments with a point or spine 20
    20. Disk with 30 to 42 segments; 4-8 mm wide; stalk
    1-3 cm high; walls of spores not calcified, but spores
    embedded in a mass of calcium carbonate. *Acicularia schenckii* p. 93
    20. Disk with 35 to 60 segments; 6-18 mm wide; stalk
    3-7 cm high; walls of spores calcified, not embedded
    in calcium carbonate. . . . . *Acetabularia crenulata* p. 92
21. Plants usually with a prominent creeping main axis (rhizome) that bears upright branches and rhizoids at intervals, always having trabeculae (internal wall struts). .22
21. Plants without a rhizomaceous main axis, without trabeculae. 31
    22. Rhizome and assimulator clearly differentiated. .23
    22. Rhizome and erect branches not distinguishable,
    plant appears as a tuft of filaments . . . . . . *Caulerpa fastigiata* p. 74
23. Upright branch a simple blade . . . . *Caulerpa prolifera* p. 77
23. Upright branches variously branched or divided. . . 24

24. Ramuli pointed at tips 25
24. Ramuli spherical, disk shaped, or elongate and ob-
    tuse at tips 29
25. Upright branches subdivided several times, appearance
    somewhat palmlike . . . . *Caulerpa paspaloides* p. 75
25. Upright branches not subdivided, not palmlike. . . . 26
    26. Blade with flat, distichous ramuli often constricted
        at the base . . . . *Caulerpa mexicana* p. 74
    26. Ramuli not flattened 27
27. Ramuli in two ranks; cylindrical and elongate. . . . . .
                                        Caulerpa sertularioides p. 77
27. Erect branches dichotomously branched with conical,
    cylindrical ramuli either ranked in two or three rows or
    randomly distributed. 28
    28. Ramuli usually in two or three rows, rhizomes
        naked . . . . *Caulerpa cupressoides* p. 74
    28. Ramuli usually randomly arranged, rhizome covered
        with ramuli similar to erect branches . . . *Caulerpa lanuginosa* p. 74
29. Erect branches 8-20 cm tall; ramuli terete, elongate, and
    usually distichous . . . . *Caulerpa ashmeadii* p. 73
29. Ramuli swollen to spherical or peltate, often in dense
    masses. 30
    30. Ramuli disklike, on short pedicels . . . . *Caulerpa peltata* p. 75
    30. Ramuli swollen, elongate to spherical . . . . *Caulerpa racemosa* p. 77
31. Plant large, ropelike, dichotomously branched; structur-
    ally of many closely united filaments. 32
31. Plant not consisting of many closely interwoven fila-
    ments to form the plant body 34
    32. Plant very dark green, rigid, densely branched with
        short internodes, dichotomous to cervicorn . . . *Codium taylori* p. 80
    32. Plant light to dark green, soft, internodes often sev-
        eral inches long, branching dichotomous. 33
33. Peripheral utricles mostly 250-300 µm in diameter, the
    end walls considerably thickened . . . . . . *Codium isthmocladum* p. 80
33. Branches often flattened below the dichotomies; utri-
    cles usually 300-400 µm in diameter, the end walls not
    thickened . . . . *Codium decorticatum* p. 80
    34. Plants coenocytic, without cross walls, appearing fil-
        amentous. 35
    34. Plant otherwise or a few large cells with septate
        haptera 41

35. Plant penetrating limestone, microscopic, filaments irregular in form, variable in diameter . . . . . *Ostreobium quecketii* p. 87
35. Plant not embedded in limestone. 36
    36. Branching dichotomous 37
    36. Branching irregular or pinnate 38
37. Filaments constricted at forks; reproductive organs ill-
defined . . . . *Boodleopsis pusilla* p. 79
37. Filaments not constricted at forks; reproductive organs
distinct, lateral, with cross walls. . . . . *Derbesia vaucheriaeformis* p. 70
    38. Plants partially embedded in soft mud, branching
    irregular . . . . *Vaucheria thuretii* p. 95
    38. Erect branches bearing distichous, radial, or secund
    ramuli 39
39. Ultimate branches radial in arrangement, main axis not
distinct from secondary branches . . . . *Bryopsis hypnoides* p. 72
39. Ultimate branches in one or two rows, main axis usually
distinct. 40
    40. Frond triangular, branchlets in two rows, in dis-
    tichous arrangement . . . . *Bryopsis plumosa* p. 73
    40. Frond linear to lanceolate, parallel in one or two
    rows, branches secund and distichous . . . . . *Bryopsis pennata* p. 72
41. Plant a flat sheet of cells 42
41. Plant not a flat blade or sheet, but some other morphology (tube, filament, etc.) 47
    42. Cells visible without magnification showing a deli-
    cate pattern of "veins" 43
    42. Cells alike and too small to be seen without magnifi-
    cation 45
43. Blades solid, without holes, showing "veins" made up
of filaments 44
43. Blades porous, having many holes and showing "veins"
made up of filaments radiating from them . .*Cystodictyon pavonium* p. 92
    44. Blades having "veins" that consist of single rows of
    filaments; plants found in shallow (0.2-10 m) water
    along coasts . . . . *Anadyomene stellata* p. 91
    44. Blades having "veins" that consist of two to five
    parallel rows of filaments; plants found in deep
    (30-60 m) water . . . . *Anadyomene menziesii* p. 90
45. Blades consist of two cells in thickness. 46
45. Blades consist of one cell in thickness. . *Monostroma oxyspermum* p. 69
    46. Blade divided into linear segments. . . . . *Ulva fasciata* p. 70

46. Blade entire, broad and irregular in shape, sometimes lobed, or linear-lanceolate and without lobes

Ulva lactuca p. 70

47. Plant elongate, branched or unbranched, a hollow tube
or a blade hollow in part, consisting of many small cells
48

47. Plant not as above. 58

48. Plants locally tubular, mostly of simple, nontubular
strands, one to three cells in cross section. . . . . .

Enteromorpha chaetomorphoides p. 67

48. Plants monostromatic and consisting of numerous
cells in cross section. 49

49. Cells more or less in longitudinal rows throughout. . 52

49. Cells not in longitudinal rows except in the youngest
parts. 50

50. Plants simple, unbranched, straight tubes. 51

50. Plants branched below, cells not in longitudinal
rows . . . . *Enteromorpha compressa* p. 67

51. Plant tubular throughout, usually inflated and irregularly constricted, often large. . . . . . . *Enteromorpha intestinalis* p. 67

51. Plant flattened, the tube adherent (as in *Ulva*) except at
the margins, where minutely hollow; lanceolate or linear-lanceolate. . . . . *Enteromorpha linza* p. 69

52. Plant unbranched, or sparingly branched at the base
only. 53

52. Plant considerably branched. 54

53. Branches, when present, at the base; cells 10-28 μm in
diameter, similar, subequal. . . . . *Enteromorpha flexuosa* p. 67

53. Branches more common although most prominent at
base, branches attenuate from base to apex; cells 8-12
μm in diameter, cell walls thin, cells not in distinct longitudinal rows. . . . . *Enteromorpha compressa* p. 67

54. Branches similar to the main axis. 57

54. Branches differing in size and shape from main axis
55

55. Plants small, slender with thornlike, abruptly tapering,
and short branches. . . . . *Enteromorpha salina* p. 69

55. Plants larger, branching of successive orders, tapering
gradually and not stubby. 56

56. Branching proliferous throughout the axes, branch-
ing not confined to the base . . . . . . *Enteromorpha prolifera* p. 69

56. Branching mostly from basal portion of axes, axes
1-2 mm broad. . . . . *Enteromorpha lingulata* p. 68

57. Plastid filling face of cell; main axes and main branches
with many long, spinelike ramuli . . . . . . *Enteromorpha ramulosa* p. 69
57. Plastid not filling face of cells, having a netlike appearance; main axes repeatedly branched without distinct
ramuli . . . . *Enteromorpha clathrata* p. 67
   58. Plant small (not over 1 mm high if filamentous) or
   larger if composed of a few large cells 59
   58. Plant filamentous, usually several cm in length,
   branched or unbranched, attached or loose 69
59. Plants solitary or a few together, consisting of large
(1.5-3.0 cm in diameter) cells with hapteral cells at the
base . . . . *Valonia ventricosa* p. 92
59. Plant otherwise, cells much smaller, usually less than 20
μm in diameter and lacking hapteral attachments . . 60
   60. Cells embedded in globose or flattened gelatinous
   matrix or sheath, separated or in clusters 61
   60. Cells not as above 62
61. Cells well separated in a gelatinous matrix, 3-7 μm in
diameter, setae lacking . . . . *Pseudotetraspora antillarum* p. 64
61. Cells in clusters in gelatinous sheaths, having distinct
setae . . . . *Diplochaete solitaria* p. 65
   62. Plant a disk or membrane of united filaments or of
   creeping, separated filaments; no distinct erect filaments 63
   62. Plants with distinct erect filaments . . . . *Stichococcus marinus* p. 66
63. Plants discoid, truly parenchymatous although fila-
ments may be separated at edges of disk 64
63. Plants usually discoid but truly filamentous throughout
although filaments may fuse laterally in central region
65
   64. Plants more than one cell in thickness, at least in
   center, edges of disk having radiating filaments, cells
   lacking pyrenoid . . . . *Ulvella lens* p. 66
   64. Plants monostromatic throughout, cells having pyrenoids and cell shape irregular, typical of littoral
   zone, producing a green zone on rocks and hard-
   woods . . . . *Protoderma marinum* p. 66
65. Filaments irregular and branching; penetrating shells or
limestone; sporangia appearing swollen . . . . . *Gomontia polyrhiza* p. 64
65. Filaments not penetrating shells or limestone 66
   66. Filaments more or less united laterally to form
   small epiphytic disks, central cells becoming en-
   larged as sporangia . . . . *Pringsheimiella scutata* p. 66

66. Filaments not united laterally. 67
67. Plants creeping, filaments branched, separate and erect;
zigzag to slightly spiraled colorless hairs arising from
cells. *Phaeophila dendroides* p. 65
67. Plants otherwise, without colorless hairs 68
68. Plants composed of spreading filaments in the wall
of hosts, with irregular cells, 3-8 μm diameter. .*Entocladia viridis* p. 65
68. Plants forming a central membrane in the walls of
Bryozoa and sponges; separate, spreading filaments
extending from edges of disk, cells 7-12 μm diam-
eter. *Entocladia flustrae* p. 65
69. Plant profusely branched 76
69. Plant unbranched, or with a few short rhizoidal
branches. 70
70. Filaments 80-300 μm in diameter 71
70. Filaments usually less than 80 μm in diameter . . 73
71. Filaments forming a loose, entangled mass. 72
71. Filaments forming a clump of attached plants, 135-250
μm in diameter. . . . . *Chaetomorpha aerea* p. 87
72. Filaments 80-175 μm in diameter . . *Chaetomorpha brachygona* p. 87
72. Filaments 180-300 μm in diameter . . . . *Chaetomorpha linum* p. 87
73. Plants small epiphytes, filaments 10-20 μm in diameter,
about 5 mm tall. . . . . *Chaetomorpha minima* p. 88
73. Plants larger epiphytes, filaments longer than 5 mm . 74
74. Filaments 40-100 μm in diameter, cells 2 to 4 diameters long, branches few to many, to several cells
long. . . . . *Rhizoclonium hookeri* p. 90
74. Filaments more slender. 75
75. Filaments 15-30 μm in diameter, with few rhizoidlike
branches. . . . .*Rhizoclonium riparium* p. 90
75. Filaments 10-14 μm in diameter, cells 3 to 7 diameters
long. . . . . *Rhizoclonium kochianum* var. *kerneri* p. 90
76. Ultimate branches usually without cross walls at the
base; filaments 180-250 μm in diameter . . . . . .
Cladophoropsis membranacea p. 92
76. Ultimate branches with cross walls at the bases . .77
77. Branching regular, dichotomous or trichotomous. . .78
77. Branching irregular, not dichotomous or trichotomous
81
78. Plants having main filament axis and one or more
netlike blades; plants not calcified. . . . . *Struvea pulcherrima* p. 92
78. Plant not consisting of a main filamentous axis and
blades. 79

79. Plants matted to prostrate, not having erect filaments; cells less than 35 µm long. 80
79. Plants not prostrate, having coarse filaments; main cells 350-400 µm long. . . . . *Cladophora longicellulata* p. 89
    80. Lower filaments more than 100 µm in diameter, cells cylindrical, color dark green. . . . . . .*Cladophora repens* p. 90
    80. Lower filaments to about 100 µm in diameter, cells of branchlets ovoid to tear shaped, color grayish green. . . . . *Cladophora frascatii* p. 88
81. Main filaments usually over 150 µm in diameter. . . 82
81. Main filaments usually less than 150 µm in diameter. 83
    82. Ultimate branches in fasciculate groups, usually over 75 µm in diameter, typically a plant of wavebeaten rocks; a deepwater form has slender ramuli 20-30 µm in diameter. . . . . *Cladophora fascicularis* p. 88
    82. Main axes and ultimate branches of about the same size, wiry dark green; typically in deep water and high salinity . . . . *Cladophora catenata* p. 88
83. Main filaments more than 60 µm in diameter, usually greater than 75 µm. 84
83. Main filaments less than 60 µm in diameter 85
    84. Main filaments distinctly angled; ultimate branches secund, over 40 µm diameter, and not clustered at branch tips; the terminal branches and cells tapering
    Cladophora gracilis p. 89
    84. Main filaments not flexuous or angled; ultimate branches strongly pectinate and clustered at branch tips; terminal cells almost bulbous and blunt tipped
    Cladophora scitula p. 90
85. Filaments soft; cells of ultimate branches usually 1 to 3 diameters long. . . . . *Cladophora delicatula* p. 88
85. Filaments relatively rigid for their size and grayish; cells of ultimate branches 3-4 diameters long or longer. . . . .
    Cladophora glaucescens p. 88

PHAEOPHYTA, SPECIES KEY

1. Plants parenchymatous, forming a hollow, lobed irregular mass . . . . *Colpomenia sinuosa* p. 110
1. Plants parenchymatous or filamentous, not forming spherical, irregular, hollow structures. 2

2. Plants filiform, soft and gelatinous in texture, constructed of filaments imbedded in a matrix of clear jelly. 3

2. Plants filamentous or otherwise but not imbedded in a gelatinous matrix 5

3. Axes solid in cross section, usually less than 1 mm wide; sporangia developed on branches borne directly upon filaments of medulla . . . . *Nemacystus howei* p. 107

3. Axes mostly hollow in cross section, usually more than 1 mm wide; sporangia formed by transformation of cortical pigmented filaments. 4

    4. Filaments of medulla separable under pressure into long regular strands; lateral filaments 20-40 μm wide. . . . . *Cladosiphon zosterae* p. 106

    4. Filaments of medulla closely united, under pressure separating into irregular strands; lateral filaments 11-15 μm wide . . . . *Cladosiphon occidentalis* p. 105

5. Plants entirely composed of free filaments, or with both filamentous and encrusting portions, or entirely encrusting 6

5. Plants neither encrusting nor composed entirely of free filaments; if filamentous in part, without encrusting portion 22

    6. Plants forming conspicuous, lobed or irregular dark brown crusts, usually on stones, shells, or epiphytic

                                                 Lobophora variegata p. 102

    6. Plants minute and inconspicuous, or other than crustose in form 7

7. Plants minute, 5 mm or less in height, often invisible without magnification or apparent only as fine fuzz or brown patches a few millimeters in diameter 8

7. Plants obvious without magnification, more than 5 mm high 13

    8. Upright filaments 1-5 mm high 9

    8. Upright filaments 0.5 mm high or less. 12

9. Base a scutate disk with terminal or lateral, longitudinally divided sporangia, usually in association with *Stictyosiphon* on sea grasses during the winter. . . . .

                                     Myriotrichia subcorymbosa p. 111

9. Base composed of ramifying filaments, usually epiphytic 10

10. Mature plant 1 mm or less, sporangia usually in clusters . . . . *Myriotrichia occidentalis* p. 110
10. Mature plant 1-5 mm, sporangia not in clusters. . 11
11. Plurilocular sporangia widest below the middle, borne
upon flexuous, irregularly branched filaments . . . *Giffordia rallsiae* p. 99
11. Plurilocular sporangia not distinctly wider below the
middle; borne mostly near the base of simple, upright
filaments . . . . *Ectocarpus elachistaeformis* p. 97
12. Elongate, hyaline paraphyses present; few erect filaments . . . . *Ascocyclus orbicularis* p. 106
12. Paraphyses absent, plurilocular sporangia 15-50 μm
long, stalked, basal layer one-cell thick . *Myrionema strangulans* p. 106
13. Filaments pluriserate, two to three cells in diameter, at
least in part; developing from a large apical cell . . . 14
13. Filaments uniseriate; growth intercalary or trichothallic
15
14. Propagulae broadly shield shaped, with short, thick
tips forming a span up to 165 μm . . . . *Sphacelaria tribuloides* p. 99
14. Propagulae slender throughout, with elongated tips
forming a span up to 450 μm . . . . . . *Sphacelaria furcigera* p. 99
15. Sporangia spherical or subspherical, 30-40 μm in diameter; numerous short branches arising at right angles to
the main axis . . . . *Acinetospora crinita* p. 96
15. Sporangia not spherical or subspherical; branching usually at an angle less than 90 degrees 16
16. Reproductive structures intercalary in the vegetative
filaments, the plastids stellate, single or double . . .
Bachelotia antillarum p. 96
16. Reproductive structures lateral or terminal on filaments; plastids band shaped or discoid 17
17. Plurilocular sporangia with acute to hair-tipped apex . .
Ectocarpus siliculosus p. 98
17. Plurilocular sporangia blunt or tapered, but not hair-
tipped 18
18. Plurilocular sporangia cone shaped, widest at the
base, maybe several in a row along a branch. . *Giffordia conifera* p. 98
18. Plurilocular sporangia not widest at the base, usu-
ally blunt at the tips 19
19. Plurilocular sporangia fusiform, long-tapering, 20-40 μm
wide and up to 250 (occasionally to 400) μm long . . .
Ectocarpus intermedius p. 97

19. Plurilocular sporangia cylindrical 20
    20. Plurilocular sporangia 10-15 μm wide, up to 250 μm
    long, often on short stalks; plastids forked or band
    shaped . . . . *Ectocarpus dasycarpus* p. 97
    20. Plurilocular sporangia usually over 15 μm wide, ses-
    sile; plastids discoid 21
21. Plurilocular sporangia 15-20 μm wide, up to 100 μm
    long; main filaments 35-45 μm wide; plant may attain
    several decimeters in length . . . . *Giffordia mitchelliae* p. 98
21. Plurilocular sporangia 25-30 μm wide, up to 250 μm
    long; main filaments 20-22 μm wide; plants soft in tex-
    ture 2-5 cm long . . . . *Giffordia indica* p. 98
    22. Plants with brushlike tufts of brown hairs, tufts
    borne on inconspicuous protuberances or slender
    branchlets found along a terete, wiry axis 23
    22. Plants lacking brushlike tufts of hairs borne on pro-
    tuberances or branchlets; hairs colorless when
    present 24
23. Pedicels of fertile branchlets up to 2 mm long; fertile
    portion 0.75-2 mm, ovoid or short cylindrical in shape;
    hair tufts usually 5 mm or less in length . . *Sporochnus pedunculatus* p. 109
23. Pedicels of fertile branchlets up to 10 mm long; fertile
    portion up to 4 mm long; hair tufts up to 10 mm long
                         Sporochnus bolleanus p. 108
    24. Plant terete throughout, or, if collapsed and flat-
    tened, hollow at least in upper portions 25
    24. Plant not terete throughout; flattened portions not
    hollow, if present 27
25. Plant terete throughout, hollow or solid and with a rela-
    tively firm texture 26
25. Plant terete, but not solid in cross section, may be solid
    near the base, texture very soft; plants flexuous and pro-
    fusely branched, mostly dichotomous, tapering from
    main axes and branches . . . . *Rosenvingea intricata* p. 111
    26. Medulla in cross section consisting of four cells sur-
    rounded by cortex of smaller cells . . *Stictyosiphon subsimplex* p. 110
    26. Medulla in cross section composed of many cells,
    may become hollow in old age . . . . . . *Stilophora rhizodes* p. 107
27. Plants mostly flattened throughout, sometimes attached
    by a terete base of blades with enrolled edges; air blad-
    ders absent 28
27. Plants differentiated into distinct terete central axes
    and lateral foliar appendages; air bladders usually

present. 38
28. Blades essentially fan shaped, with enrolled edges;
    four to eight cells thick . . . . *Padina vickersiae* p. 103
28. Blades fan shaped or otherwise; edges not enrolled
    29
29. Blades with distinct midrib, blades two cells thick. . . .
    Dictyopteris delicatula p. 100
29. Midrib absent or indistinct. 30
    30. Blades essentially fan shaped . . . . *Lobophora variegata* p. 102
    30. Blades strap shaped or lanceolate in shape 31
31. Blades growing from an apical row of cells; in cross
    section with a medulla eight cells thick . . *Spatoglossum schroederi* p. 104
31. Blades growing from a single apical cell; in cross section
    with a medulla usually one cell thick (or one to four
    cells thick). 32
    32. Blades with a medulla one cell thick throughout . 33
    32. Blades with medulla more than one cell thick at the
        margins, especially in the older, basal portions. . . .
        Dilophus alternans p. 102
33. Branching alternate; margins dentate or erose-dentate
    Dictyota dentata p. 100
33. Branching dichotomous, cervicorn or irregular. . . . 34
    34. Blades 1 mm wide or less throughout; rarely wider
        Dictyota linearis p. 102
    34. Blades mostly or entirely more than 1 mm wide. 35
35. Branching cervicorn to irregular, with reduced member
    often spurlike; branches typically 2.5 mm or less in
    width . . . . *Dictyota cervicornis* p. 100
35. Branching essentially dichotomous 36
    36. Ultimate branchlets filiform, to 0.5 mm wide,
        abruptly reduced from broader branches . . *Dictyota divaricata* p. 101
    36. Ultimate branches wider or, if 0.5 mm wide, slender
        throughout or gradually reduced in size toward the
        tips, distance between forkings 5 to 10 times the
        branch width, rarely up to 20 times branch width. .
        37
37. Thallus margin aculeate to dentate in regular fashion. . .
    Dictyota ciliolata p. 100
37. Thallus margin entire at best slightly proliferous. . . . .
    Dictyota dichotoma p. 101
    38. Plants pelagic, never attached or forming a holdfast;
        cryptostomata lacking or greatly reduced; repro-
        ductive structures unknown 39

RHODOPHYTA, SPECIES KEY

2. Plants erect, filamentous, branched or unbranched
   3
3. Plant consisting of simple, unbranched filaments with
   lobed to short rhizoidal cells . . . . *Erythrotrichia carnea* p. 117
3. Plants erect, branching pseudodichotomous 4
   4. Cells elongate, grayish green . . . . *Asterocystis ramosa* p. 117
   4. Cells cuboidal to squat, rose red . . . . . *Goniotrichum alsidii* p. 118
5. Plant impregnated with lime, sometimes stony to touch
   and whitened in older plants 6
5. Plants not stony and not impregnated with lime . . . 18
   6. Plant forming a crust; if erect, then not jointed . . 7
   6. Plant erect, flexible, or, if heavily calcified, then
      jointed; branches may arise from a crustose base . 12
7. Plant consisting of one to three layers of cells, delicately
   crustose 8
7. Plant usually more than three cell layers thick, crustose
   or irregularly branched 10
   8. Plant monostromatic (one cell layer thick) . . *Fosliella atlantica* p. 123
   8. Plant one to three cell layers thick, variable 9
9. Distinct large cells present; heterocysts in surface view,
   cells varying in size from 7-10 μm to 15-18 μm . . *Fosliella farinosa* p. 123
9. No large hair cells present; structural cells in surface
   view, 6-7 μm broad . . . . *Fosliella lejolisii* p. 124
   10. Vegetative tissue uniform, no heterocysts present . 11
   10. Vegetative tissue containing heterocysts. . . . . . .
      .... *Goniolithon decutescens* p. 124
11. Plant crustose, conforming to substrate and often super-
    imposed but not strongly adherent . *Lithothamnium syntrophicum* p. 126
11. Plant with knoblike irregular branches, not conforming
    to substrate, becoming free . . . . . . *Lithothamnium occidentale* p. 126
    12. Plant soft, with light calcification in outer cortex . .
       .... *Galaxaura obtusata* p. 119
    12. Thallus stony to touch, heavily calcified 13
13. Conceptacles at tips of terminal segments, later over-
    passed by growth 15
13. Conceptacles scattered over the surface (surface cavities
    to about 500 μm in diameter) 14
    14. Segments distally and unequally forked, coarse,
       ends without swellings . . . . . . *Amphiroa rigida* var. *antillana* p. 122
    14. Branching dichotomous, segments themselves rarely
       forked, ends with padlike swellings . . . . *Amphiroa fragilissima* p. 122
15. Branching lateral . . . . *Corallina cubensis* p. 122
15. Branching dichotomous 16

16. Plants minute, tufts about 3 mm in diameter . . . *Jania pumila* p. 125
16. Plants much larger, tufts usually 0.4 to 5.0 cm in
height 17
17. Segments more than 100 µm in diameter in older (lower)
portions of plant, 2 to 4 diameters long; ultimate
branchlets slender . . . . *Jania adhaerens* p. 124
17. Segments less than 100 µm in diameter, 4 to 6 diam-
eters long, entire plant appearing delicate . . . . . *Jania capillacea* p. 125
18. Plant with erect branches or blades, sometimes a
crustose attachment 20
18. Plant crustose, without erect branches 19
19. Tetrasporangia in sunken conceptacles; lacking paraphyses; plant a thin crust (to 1 mm) . . . *Hildenbrandia prototypus* p. 122
19. Tetrasporangia in raised nemathecia with paraphyses;
plant a lobed crust (2-10 mm); plant slightly calcified . .
Peyssonnelia rubra p. 122
20. Plant more than one cell in thickness (pluriseriate or
multiseriate), at least at the nodes 33
20. Plant uniseriate; a branched or unbranched filament
21
21. Plant small but macroscopic, cells with numerous discoid plastids 26
21. Plant microscopic (to about 3 mm tall); cells containing
single or few simple, lobed, or stellate-shaped plastids . .
22
22. Plastids parietal, simple or lobed 23
22. Plastids axial and stellate . . . . *Kylinia crassipes* p. 119
23. Base consisting of a single cell, modified from the original germinating spore . . . . *Acrochaetium sargassi* p. 118
23. Base consisting of creeping filaments 24
24. Basal filaments penetrating outer portion of host
(usually *Avrainvillea*), monosporangia single or a
few in series . . . . *Acrochaetium avrainvilleae* p. 118
24. Basal filaments entirely external to the host . . . 25
25. Basal filaments forming a subparenchymatous disk, not
clearly filamentous . . . . *Acrochaetium thuretii* p. 118
25. Base evidently filamentous although may be congested
in center, sporangia in long series on branches . . . . . .
Acrochaetium seriatum p. 118
26. Plant appearing bushy, central uniaxial filament
covered with short, determinate branchlets . . *Crouania attenuata* p. 146
26. Plant with open branching, lacking determinate
branchlets 27

27. Branching primarily opposite or whorled . . . .
27. Branching alternate to dichotomous 28
   28. Lateral branchlets generally smaller than the main
    axis; colorless hairs absent 30
   28. Main axis and branchlets uniform in size, with ter-
    minal colorless hairs 29
29. Cells swollen to nearly spherical . . . . *Griffithsia globulifera* p. 146
29. Cells not notably swollen; cylindrical . . . . *Griffithsia tenuis* p. 146
   30. Asexual reproduction by radiating chains of thick-
    walled spores (seirospores), vegetative cells uni-
    nucleate . . . . *Seirospora occidentalis* p. 147
   30. Asexual reproduction by tetraspores, vegetative
    cells plurinucleate 31
31. Older portions (base) of thallus corticated by rhizoidal
   filaments . . . . *Callithamnion roseum* p. 143
31. Plants uncorticated, or with a few rhizoidal filaments
   not covering main axis 32
   32. Main filaments about 160 μm in diameter, plants
    2-4 cm . . . . *Callithamnion cordatum* p. 143
   32. Main filaments 40-100 μm in diameter, plants 3-12
    cm . . . . *Callithamnion byssoides* p. 143
33. Thallus having a polysiphonous axis visible throughout
   the plant when young or visible in a cross section . . 34
33. Thallus not visibly polysiphonous in construction . . 72
   34. Branches and main axis at least at the growing tips
    covered with uniseriate, pigmented (cells containing
    plastids) filaments 70
   34. Branches and main axis not so covered although the
    final branchlets may terminate in uniseriate fila-
    ments 35
35. Terminal branchlets uniseriate, at least the last few segments; all cells pigmented (containing plastids) . . . 36
35. Terminal branchlets corticated or polysiphonous, not
   ending in uniseriate segments, although colorless uniseriate hairs may be present 43
   36. Holdfast consisting of ordinary branches (or rhi-
    zoids) arising in the normal branch position . . . 37
   36. Holdfast forming special outgrowth from the main
    axis, formed from the lower side and at a fork and
    not arising in a normal branch position 40
37. Thallus radially organized and consisting of four to
   seven pericentral cells 38

37. Thallus dorsiventrally organized and containing 4 to 12
    pericentral cells 39
    38. Pericentral cells five to seven . . . . *Brongniartella mucronata* p. 154
    38. Pericentral cells four . . . . *Murrayella periclados* p. 160
39. Branchlets altogether uniseriate . . . . . . *Bostrychia moritziana* p. 153
39. Final segments of branchlets uniseriate, basal portions
    polysiphonous . . . . *Bostrychia radicans* p. 153
    40. Branch systems all having similar growth potential,
        not differentiated into long and short branches. . .
                            Bostrychia scorpioides var. *montagnei* p. 153
    40. Branch system of long, axial branches and short,
        one- to three-pinnate, lateral branches; plant appear-
        ing feathery in structure 41
41. Ultimate branchlets altogether monosiphonous . .*Bostrychia tenella* p. 154
41. Ultimate branchlets polysiphonous, at best only last
    few segments monosiphonous 42
    42. Cortication rhizoidal in nature, the branchlets poly-
        siphonous . . . . *Bostrychia calliptera* p. 152
    42. Cortication parenchymatous, ultimate branchlets
        polysiphonous at base, only last few segments
        monosiphonous . . . . *Bostrychia binderi* p. 152
43. Plant with creeping, dorsiventral axis, sometimes with
    erect filaments and branches 66
43. Plant erect, radially organized 44
    44. Axis of plant heavily corticated, pericentral cells
        visible only in cross section, even at growing tips .45
    44. Axis of plant corticated or not, but pericentral cells
        at least visible without sectioning in young growing
        tips 52
45. Apical cells sunken at tips of branches; branchlets
    appearing blunt 46
45. Apical cells not in pits, exposed at tips of branches or
    hidden by colorless hairs 51
    46. Pericentral cells with conspicuous thickenings in
        end walls, banding clearly visible through cortex,
        especially under a dissecting microscope . . *Chondria collinsiana* p. 154
    46. Pericentral cells lacking conspicuous end walls, hav-
        ing no unusual banding when seen under a dissect-
        ing microscope 47
47. Plants lax; sprawling, flexuous main axis 0.78-1.25 mm
    in diameter . . . . *Chondria cnicophylla* p. 154
47. Plants bushy and erect 48

48. Plants large, bushy, strongly staining the paper
when dried, branchlet tips bearing conspicuous tufts
of trichoblasts . . . . *Chondria dasyphylla* p. 155
48. Plants not strongly staining paper, branchlet tips
lacking dense tufts of trichoblasts, although hairs
may be present 49
49. Branchlets contracted or constricted at base 50
49. Branchlets not constricted or contracted at base, trichoblasts inconspicuous or absent . . . . *Chondria floridana* p. 155
50. Plants erect and slender, axes 0.22-0.32 mm in
diameter . . . . *Chondria leptacremon* p. 155
50. Plants pyramidal and bushy, axes 1.0-2.5 mm in
diameter . . . . *Chondria sedifolia* p. 155
51. Plants large, brownish purple, branches slender and distinctly different from the main axis. . . . . *Chondria littoralis* p. 155
51. Plants coarse, branchlets spindle shaped, showing regu-
lar gradation in size compared to the main axis .*Chondria tenuissima* p. 156
52. Pericentral cells three, plants uncorticated . . . . .
Falkenbergia hillebrandii p. 157
52. Pericentral cells four or more 53
53. Pericentral cells four 54
53. Pericentral cells five or more 62
54. Main branches heavily corticated, bearing determinate branchlets 2-5 mm long . . . . *Polysiphonia ramentacea* p. 162
54. Main branches not corticated or slightly corticated
in older (basal) regions 55
55. Apical cell exposed, trichoblasts few or absent . . . . .
Polysiphonia subtilissima p. 162
55. Apical cell usually obscured by cluster of trichoblasts. 56
56. Branching subdichotomous, most evident in older
(basal) portions 60
56. Branching alternate in upper parts of plant 57
57. Branches arising in the axils of trichoblasts . . . .
Polysiphonia havanensis p. 161
57. Branches replacing trichoblasts 58
58. Segments to 90 μm in diameter, large apical cell
apparent in spite of trichoblasts . . . *Polysiphonia subtilissima* p. 162
58. Segments more than 100 μm in diameter, apical cell
usually hidden by trichoblasts 59
59. Older segments 100-150 μm in diameter, 1.5 to 2.0
diameters long . . . . *Polysiphonia macrocarpa* p. 162
59. Older segments 200-300 μm in diameter, shorter than

broad . . . . *Polysiphonia ferulacea* p. 161
60. Main axis having spinelike, determinate branchlets;
plants larger than 2.5 cm 61
60. Main axis without spinelike, determinate branchlets;
plants small (to 2.5 cm tall) epiphytic on sea grasses
with discoid base . . . . *Polysiphonia gorgoniae* p. 161
61. Spinelike, determinate branches closely set, reaching 1
mm in length; plants coarse below, branching in wide
angles . . . . *Polysiphonia echinata* p. 161
61. Spines more scattered, 2 mm in length; plants more
delicate and branching in clusters . . . . *Polysiphonia hapalacantha* p. 161
62. Pericentral cells six, rarely five to eight . *Polysiphonia denudata* p. 160
62. Pericentral cells eight or more 63
63. Pericentral cells eight, branchlets distinctively short,
closely covering the axis . . . . *Bryocladia cuspidata* p. 154
63. Pericentral cells 8 to 24, branchlets not distinct or
closely covering the axis 64
64. Pericentral cells 8 to 12; if more, then plants not
dark and coarse in appearance 65
64. Pericentral cells 12 to 24; plants dark and coarse in
branching . . . . *Polysiphonia opaca* p. 162
65. Pericentral cells 8 to 12; plants to 8 cm tall; segments 1
to 2 diameters in length . . . . *Polysiphonia tepida* p. 162
65. Pericentral cells 10 to 15; plants small, creeping, soft,
with short, erect branches; segments shorter than broad
Polysiphonia howei p. 161
66. Erect branches of two types: indeterminate, polysiphonous (four pericentral cells) branches and
determinate, filiform, flat branchlets . . . . . *Taenioma nanum* p. 151
66. Erect branches not of two types, similar to rhizome
67
67. Erect branchlets indeterminate; sparingly branched;
trichoblasts forming a crest on dorsal side of branchlet
68
67. Erect branches determinate and unbranched, in regular
sequence from a creeping axis; trichoblasts radially arranged 69
68. Rhizoids slender, pericentral cells 6 to 12. *Lophosiphonia cristata* p. 160
68. Rhizoids bulbous, pericentral cells four . . . .
Lophosiphonia saccorhiza p. 160
69. Each segment of main axis with branchlet, long branch,
or branch rudiment . . . . *Herposiphonia tenella* p. 158

69. Some segments of main axes without branches or
branchlets . . . . *Herposiphonia secunda* p. 157
70. Plants having five pericentral cells in main axis . . .
Wrightiella tumanowiczi p. 163
70. Plants having four pericentral cells in main axis. .71
71. Uniseriate branchlets in spiral arrangement, main axis
extensively corticated . . . . *Dasya corymbifera* p. 164
71. Uniseriate branchlets in random patterns, primary
branching sparse . . . . *Dasya pedicellata* p. 165
72. Thallus solid or interior consisting of dense or loose
arrangements of filaments or a single row of cells
(uniaxial) with cortication 77
72. Thallus hollow throughout or hollow in part or hollow and containing septa 73
73. Thallus consisting of solid, wiry, and slender axis supporting hollow vesicular branchlets . . . . *Botryocladia occidentalis* p. 141
73. Thallus entirely hollow or with septa 74
74. Thallus hollow throughout, no septations, but
branchlets locally constricted . . . *Chrysymenia enteromorpha* p. 141
74. Thallus hollow with true septations, at least at the
base of the branchlets 75
75. Plants hollow, the branches constricted and septate
only at the bases 76
75. Plants hollow and divided into visible segments by
transverse septa, usually constricted at sites of septa . . .
Champia parvula p. 139
76. Plant procumbent, at least the main axis and base,
erect branchlets simple . . . . *Lomentaria rawitscheri* p. 140
76. Plant erect, base simple, not creeping; plants bushy,
well branched . . . . *Lomentaria baileyana* p. 140
77. Plant in cross section parenchymatous, lacking cluster
of central filamentous cells 97
77. Plant in cross section multiaxial, that is, revealing a
loose or dense filamentous (multiaxial) medulla . . . 78
78. Plant terete to compressed in cross section, not
bladelike 79
78. Plant bladelike to foliaceous, often with a number
of blades 91
79. Plant consisting of compressed oval to spindle-shaped
segments, usually found creeping on mangrove roots . .
Catenella repens p. 133
79. Plant otherwise, not segmented. 80

80. Plant wiry, axes of small diameter (0.5-5 mm) . . 81
80. Plant cartilaginous, gelatinous but not wiry, axes of
    large diameter (5-15+mm) 85
81. Medullary filaments of two types in older parts of
    plants: an extremely thick-walled filament that in cross
    section appears lens shaped (rhizines), and thin-walled
    filaments that do not appear lens shaped in cross sec-
    tion. 83
81. Medullary filaments thin-walled and of one type . . .82
    82. Branch tips with single apical cell (thallus uniaxial
        in construction) . . . . *Gelidiella acerosa* p. 119
    82. Branch tips with numerous apical cells (thallus mul-
        tiaxial in construction) . . . . *Wurdemannia miniata* p. 120
83. Rhizines chiefly in the central medulla . . . . *Pterocladia americana* p. 120
83. Rhizines chiefly in subcortical areas 84
    84. Plants to 0.5 cm tall, with flattened, erect branches
                                        Gelidium pusillum p. 120
    84. Plants 2-5 cm tall, with cylindrical, almost filiform,
        erect branches . . . . *Gelidium crinale* p. 120
85. Plants very soft and gelatinous; dichotomously
    branched; medulla appearing vacant with occasional
    filaments .86
85. Plant cartilaginous to firm; not dichotomously
    branched; medulla loose to compact; plants not easily
    squashed .87
    86. Cortex of polyhedral cells, formed by dichoto-
        mously branched medullary filaments . . . *Scinaia complanata* p. 119
    86. Cortex of large, inner cells surrounded by small
        cells; medulla of slender filaments embedded in jelly
                                        Halymenia agardhii p. 126
87. Medulla composed of a densely compacted bundle of
    filaments, plants cartilaginous 88
87. Medulla composed of loosely arranged filaments, plants
    firm but not cartilaginous 89
    88. Branches having whorls of spines or nodules. . . . .
                                        Eucheuma isiforme ꜰ p. 135
    88. Branches smooth, lacking or having few spines . . .
                                        Eucheuma nudum p. 137
89. Branching alternate or radial, division terete . . . *Agardhiella tenera* p. 135
89. Branching distichous, main axis compressed to flattened
                                        90
    90. Main axis at best compressed, branching in a zigzag

pattern; plant fleshy, with ultimate branches spinelike . . . . *Eucheuma acanthocladum* p. 135

90. Main axis flattened to compressed, branching regular; plant soft, gelatinous; branches not spinelike . .
Agardhiella ramosissima p. 133

91. Blades firm; medulla composed of filaments that anastomose; filaments abundant in cross section, not in
jelly 92

91. Blades soft to gelatinous; without anastomosing filaments; medulla composed of filaments separated in soft
jelly 95

92. Blades strap shaped, marginally pinnately branched,
also having proliferations from the surface . . *Grateloupia filicina* p. 126

92. Blades lobed, cleft or irregularly branched 93

93. Blades lobed, irregularly to palmately cleft; sporangia
zonate . . . . *Meristotheca floridana* p. 138

93. Blades simple, orbicular to broadly lobed, sporangia
tetrapartite 94

94. Plants composed of small (0.5-15 mm in diameter),
simple-lobed blades . . . . *Kallymenia limminghii* p. 128

94. Plants composed of larger blades (1-2 dm in diameter) with a regular pattern of perforations . . . . .
Kallymenia perforata p. 128

95. Flattened thallus irregularly lobed or proliferous, blades
to 6 dm wide, subcortical stellate cells common. . . . .
Halymenia gelinaria p. 127

95. Flattened thallus pinnately lobed or branched 96

96. Branching compounded two to three times with
ultimate marginal branchlets becoming filiform . . .
Halymenia floresia p. 126

96. Branching compounded one to two times, the
broad, ultimate segments serrated . . *Halymenia pseudofloresia* p. 127

97. Mature portions of plant parenchymatous or, if uniaxial, the central filament (cell row) obscured except at
growing tips 108

97. Mature portion clearly uniaxial, the central cell row not
obscured, with continuous or nodal cortication . . . 98

98. Ultimate branchlets (ramuli) corticated in bands
only, major indeterminate branches fully corticated
Spyridia filamentosa p. 148

98. Cortication progressive and similar on all branches
99

99. Cortication complete, cells in longitudinal rows, nodes
with whorls of spines . . . . *Centroceras clavulatum* p. 143
99. Cortication incomplete, usually in bands, cells not in
longitudinal rows, no spines at nodes 100
   100. Nodal bands of mature axes with more than two
   rows of cells 102
   100. Nodal bands with only one to two transverse rows
   of cells 101
101. Nodes of erect filaments about 20-35 μm in diameter;
usually epiphytic on *Codium* . . . . *Ceramium codii* p. 145
101. Nodes about 50 μm in diameter; habit various . . . . . .
              Ceramium fastigiatum f. *flaccida* p. 145
   102. Branching alternate and close; distal parts frondlike . . . . *Ceramium floridanum* p. 145
   102. Branching dichotomous, loose; no frondlike appearance 103
103. Lower one or two cells of nodal bands regularly rectangular, broader than long . . . . *Ceramium byssoideum* p. 143
103. Lower one or two cells of the nodal bands longer than
broad 104
   104. Cortical bands in ultimate branches confluent with
   one another, although limited to nodes in older portions of plant . . . . *Ceramium corniculatum* p. 145
   104. Cortication not confluent in any region . . . 105
105. Nodal cells in distinct transverse series 107
105. Nodal cells irregular 106
   106. Diameter of nodes of older portions 210-330 μm;
   sporangia immersed by nodal cortication . . *Ceramium strictum* p. 146
   106. Diameter of nodes of older portions 60-155 μm;
   sporangia naked and emergent . . . . . . *Ceramium fastigiatum* p. 145
107. Nodal cortication consisting of one row of small and
large cells; small cortical cells sometimes subdivided
in older portions; nodes 50 μm in diameter . . . . .
              Ceramium fastigiatum f. *flaccida* p. 145
107. Nodal cortication of more than two rows of cells; nodes
to 120 μm in diameter . . . . *Ceramium subtile* p. 146
   108. Plant terete or cylindrical, at best compressed . 116
   108. Plant flattened or foliaceous, at least in older portions 109
109. Plant a single flattened blade, up to 10 cm wide, with a
prominent midrib . . . . *Grinnellia americana* p. 150
109. Plant otherwise, midrib either absent or, if present,
plant small and midrib inconspicuous 110

110. Blades peltate, stipe rarely marginal . . . . . *Fauchea peltata* p. 142
110. Blades pinnately or dichotomously branched . 111
111. Branches without evident midrib, main branches 4-5 mm or more broad. 112
111. Branches with midrib in small flattened blades, 1-3 mm wide 115
   112. Plants pinnately to alternately branched, segments fusiform to flattened . . . . *Gracilaria foliifera* p. 130
   112. Plants dichotomously or subpalmately branched, segments flattened. 113
113. Plants dichotomously branched but lacking proliferous margins. 114
113. Plants dichotomously branched to subpalmately branched, having numerous small marginal proliferations . . . .
                                    Cryptopleura fimbriata p. 150
   114. Plants consisting of dichotomously branched, strap-shaped segments with oblong to rounded tips, thick and leathery. . . . . *Gracilaria mammillaris* p. 131
   114. Plants dichotomously branched, fan shaped in appearance, segments thin and soft . . . . *Fauchea hassleri* p. 142
115. Plant 4-5 cm high, with dichotomous branches mostly in one plant; purple colored; creeping littoral species; plant forming flat dichotomies, branches constricted at forkings. *Caloglossa leprieurii* p. 149
115. Plants larger (to 10 cm); rose to green colored; bushy, branching in all planes, branches arising from midrib of blades. . . . . *Hypoglossum tenuifolium* p. 151
   116. Axis clothed with uniseriate, pigmented filaments. .
                                    Dasyopsis antillarum p. 165
   116. Axis lacking uniseriate, pigmented filaments. . 117
117. Apical cells sunken in pits at tips of branches. . . . 118
117. Apical cells not sunken in pits at tips of branches. . 123
   118. Branching usually widely divergent, cortical cells in apical portions having small projections (mammillate) at tip of cell . . . . *Laurencia gemmifera* p. 158
   118. Branching usually not divergent, cortical cells of apical portion lacking mammillate projections. . 119
119. Walls of axial cells of main branchlets thickened and visible as transverse bands through the thallus, branchlets 0.20 to 0.45 mm in diameter. . . . . . *Laurencia microcladia* p. 158
119. No visible transverse banding in main axis, branchlets larger 120
   120. Ultimate branchlets blunted, truncate and appear-

ing wartlike; a closely branched plant . . . *Laurencia papillosa* p. 159

120. Branchlets more elongate, not closely set; main
     axis bearing branchlets on flattened margins. . 121

121. Plants stiff, cartilaginous, ultimate branches a peg or
     stubby outgrowth . . . .*Laurencia poitei* p. 159

121. Plants soft, main axis 0.5-1.5 mm in diameter; ultimate
     branchlets club shaped to obovoid. 122

   122. Plants erect, with well-defined main axis; short,
        truncate branchlets. . . . . *Laurencia obtusa* p. 158

   122. Plants entangled, without evident main axes;
        branchlets subcylindrical to slightly clavate. . . . .
                                        Laurencia intricata p. 158

123. Main axes covered by short, stiffly filamentous, but
     thinly corticated branchlets. . . . .*Digenia simplex* p. 156

123. Main axis not so covered; branchlets, if present, short
     and heavily corticated. 124

   124. Axes with short, spinelike branchlets on short,
        spurlike branches. 125

   124. Main axes lacking both spines and short, spurlike
        branches. 127

125. Axis with uniaxial, central filament visible at least at
     tips and with stellate or cornute spines. . . . . . *Hypnea cornuta* p. 139

125. Axis not uniaxial and without cornute spines. . . . 126

   126. Terminal branches crowded; short spines present
        on indeterminate branches (main axes) as well as on
        the specialized, blunt, determinate branches. . . . .
                                    Acanthophora muscoides p. 151

   126. Terminal branches loosely branched, ultimate divi-
        sions often long; short spines only on the blunt de-
        terminate branches. . . . . *Acanthophora spicifera* p. 151

127. Multiaxial construction; branch tips with numerous
     apical cells, no central filament present in young
     branches. .128

127. Uniaxial construction; branch tip with one apical cell,
     axes showing a single central filament near the tips . 137

   128. Plants terete throughout. 131

   128. Main axis and some branches usually flattened. 129

129. Branching primarily dichotomous. 130

129. Branching primarily pinnate; main axis compressed and
     irregularly branched. . . . . *Gracilaria cervicornis* p. 129

   130. Branching alternate to subdichotomous; divisions
        erect; main axis flattened. *Gracilaria foliifera* var. *angustissima* p. 131

130. Branching secund and incurved, flattening limited
  to regions of branching. . . . .*Gracilaria compressa* p. 129
131. Branches cylindrical and constricted at the base, medullary cells large (400-1,000 μm in diameter) . . . 132
131. Branches cylindrical or slightly compressed and not
  constricted at the base, medullary cells small. . . . 133
  132. Plants having one to two degrees of branching at
    most; medullary cells 400-500 μm in diameter;
    plants rather cartilaginous. . . . .*Gracilaria cylindrica* p. 129
  132. Plants having three or more degrees of branching;
    medullary cells visible from outside, 600-900 μm in
    diameter; plants soft. . . . .*Gracilaria blodgetti* p. 128
133. Plants coarse, cartilaginous, moderately branched; distal
  branching secund. . . . .*Gracilaria debilis* p. 129
133. Plants slender, branching abundant. 134
  134. Main branching dichotomous, smaller plants con-
    gested; cartilaginous. . . . . *Gracilaria damaecornis* p. 129
  134. Main branching alternate, at least at the top. . 135
135. Plants firm, wiry in texture, branchlets secund on lesser
  branchlets. . . . . *Gracilaria armata* p. 128
135. Plants soft to firm in texture, branching radial and alternate, dichotomous below. 136
  136. Cortex of two to three layers of small cells. . . . .
    Gracilaria verrucosa p. 132
  136. Cortex of four to eight layers of small cells in rows. .
    Gracilariopsis sjoestedtii p. 132
137. Plant with occasional, thickened and recurved (crosier)
  branch tips. . . . . *Hypnea musciformis* p. 138
137. Branching without recurved branch tips. 138
  138. Axes with stellate or cornute spines, at least in
    older portions distinct . . . . *Hypnea cornuta* p. 139
  138. Stems smooth, or, if spiny, spines simple. . . 139
139. Main axes to 15 cm long, with abundant simple spines
  of coarse texture; plant forming loose, soft clumps. . .
    Hypnea cervicornis p. 139
139. Main axes to 4-5 cm long; plants forming dense mats in
  wave-washed areas; slender (to 1 mm in diameter); spines
  sparse, sometimes developing into branchlets . . *Hypnea spinella* p. 139

4

Cyanophyta

SPECIES DESCRIPTIONS

Because of the major taxonomic revision of genera and species of the Chroococcales and Oscillatoriaceae in the division Cyanophyta (Drouet, 1968; Drouet and Daily, 1956), synonyms are included for taxa so that the investigator may refer to previous publications. (Tilden, 1910; Humm, unpublished manuscript; Desikachary, 1959).

ORDER CHROOCOCCALES

Family Chroococcaceae

Agmenellum quadruplicatum Brébisson
  Synonym: *Merismopedia punctata* Meyen
  Drouet and Daily (1956), p. 86; Desikachary (1959), p. 155.
  Plants a colony of cells arranged in a sheet in ordered rows, cells 1-4 μm in diameter, globular.
  A number of specimens collected on shells in the littoral zone at Point of Rocks.

Agmenellum thermale (Kützing) Drouet *et* Daily
  Synonym: *Merismopedia elegans* A. Braun
  Figs. 4, 5
  Humm (unpublished), p. 22; Drouet and Daily (1956), p. 89; Desikachary (1959), p. 156.

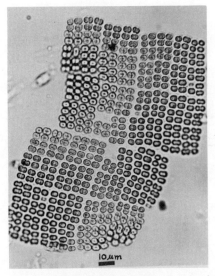 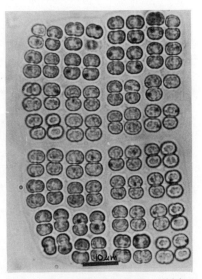

Fig. 4. *Agmenellum thermale,* a general view of a large colonial plate.

Fig. 5. *Agmenellum thermale*, a higher magnification showing cellular arrangement and the faint colonial matrix.

Plant a colony of cells in regular rows forming a sheet. Cells ovoid, light green, 4-6 μm in diameter and up to 10 μm in length. Colonies small with transparent matrix.

Collected on rocks and in sand at Ana Maria Island, on outer jetties, and at Venice Jetty.

Anacystis aeruginosa Drouet *et* Daily
   Fig. 6
   Humm (unpublished), p. 21; Drouet and Daily (1956), p. 76.
   Cells spherical to subspherical, 5-13 μm in diameter with a thickened sheath that may be colored. Cells irregularly arranged within the sheath and forming irregular colonies.
   Collected on shell and coquina rock (Point of Rocks).

Anacystis dimidiata Drouet *et* Daily
   Figs. 7, 8
   Humm (unpublished), p. 21; Drouet and Daily (1956), p. 70.
   Microscopic colony of 2 to 4 up to 12 cells, but not forming large colonies. Cell 10-16 μm in diameter with flattened walls, somewhat cubically arranged. Sheath hyaline and homogeneous in texture.

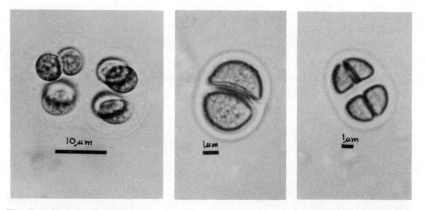

Fig. 6. (left) *Anacystis aeruginosa*, consists of eight daughter cells seen just after division. Fig. 7. (center) *Anacystis dimidata*, two-cell stage. Fig. 8. (right) *Anacystis dimidata*, four-cell stage.

Collected in sand samples from various beaches (Siesta Key; Ana Maria Island).

Anacystis montana (Lightfoot) Drouet *et* Daily
    Synonyms: *Chroococcus minutus* (Kützing) Nageli
                Chroococcus minor (Kützing) Nageli
    Fig. 9
    Humm (unpublished), p. 20; Drouet and Daily (1956), p. 45; Desikachary (1959), pp. 103, 105.
    Plants forming colonies, planktonic or forming an irregular layer on sand. Cells spherical to ellipsoidal, 1.5-5 μm in diameter, tightly packed in groups of up to four.
    Collected in littoral zone in rock and sand at Point of Rocks and Ana Maria Island.

Coccochloris elabens (Brébisson) Drouet *et* Daily
    Synonym: *Microcystis elabens* (Brébisson) Kützing
    Fig. 10
    Humm (unpublished), p. 19; Drouet and Daily (1956), p. 28; Desikachary (1959), p. 97.
    Colony of cells embedded in a gelatinous matrix, the cells spherical to elliptical in shape, measuring 3-4 up to 6 μm in diameter and 6-8 μm in length. Coloration olive to blue green.
    Common in littoral zone in sand (Point of Rocks) and on shells (Venice Jetty).

Gomphosphaeria aponina Kützing

Humm (unpublished), p. 24; Drouet and Daily (1956), p. 98.

Plants forming spherical colonies in gelatinous, hyaline matrix, the cells usually radially oriented, the cells pyriform to heart shaped, 4-15 μm in diameter and about 6 μm in length.

Collected as epiphytes on *Cladophora* and also found on rocks at the outer jetties, Ana Maria Island.

Johannesbaptistia pellucida Taylor *et* Drouet

Drouet and Daily (1956), p. 85.

Colonies of discoid cells before division, arranged in a linear series within a gelatinous matrix. Cell division occurring in a plane along the cell at right angles to the cell axis, thus forming a pseudotrichome. Cells 3.6-8.0 μm in diameter.

Collected in salt marshes, at the mouth of Homosassa River, growing with other blue-greens at the bases of the salt grass *Spartina*.

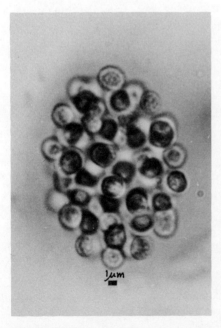

Fig. 9. *Anacystis montana*, a small, dense colony.

Fig. 10. *Coccochloris elabens*, a two-celled colony.

Family Chamaesiphonaceae

Entophysalis conferta (Kützing) Drouet *et* Daily
Synonym: *Dermocarpa olivacea* (Reinsch) Tilden
Humm (unpublished), p. 26; Tilden (1910), p. 55; Drouet and Daily
(1956), p. 111.

Plants epiphytic to endophytic; starting as a single cell then producing a
cushion; surface cells 3-7 μm in diameter, mainly angular to spherical, blue
green in color. The sheath distinct and hyaline.

Epiphytic on *Cladophora* filaments, pneumatophores of the black mangrove at Sarasota Pass (east side of Ana Maria Island), and on limestone at
Gull Rock (mouth of Homosassa River).

Entophysalis deusta (Meneghini) Drouet *et* Daily
Synonyms: *Hyella caespitosa* Bornet *et* Flahault
Pleurocapsa fuliginosa Hauck
Humm (unpublished), p. 25; Tilden (1910), p. 48; Drouet and Daily
(1956), p. 103; Desikachary (1959), p. 183.

Plants a single layer of columnar cells, producing a strata or cushion over
rocks and shells. The cells blue green, showing some false branching up to 7
μm in diameter and 20 μm long to longer. Sheath thick and colorless.

Collected on granite rocks at Venice Jetty and Point of Rocks and from
limestone at the mouth of Homosassa River.

ORDER NOSTOCALES

Family Oscillatoriaceae

Arthrospira brevis (Kützing) Drouet
Synonym: *Oscillatoria brevis* Kützing
Tilden (1910), p. 79; Desikachary (1959), p. 241; Drouet (1968), p. 219.

Trichomes 2-10 μm in diameter, having cross walls; cross walls lined with
granular, dense protoplasm; filaments straight or coiled. Terminal cell conical,
the outer wall of tip cell thin, cells of trichome up to six times as short as
broad, sheath material hyaline or absent.

Collected in salt marsh, mouth of Homosassa River.

Microcoleus lyngbyaceus (Kützing) Crouan
Synonyms: *Oscillatoria corallinae* (Kützing) Gomont
Oscillatoria curviceps Agardh *ex* Gomont

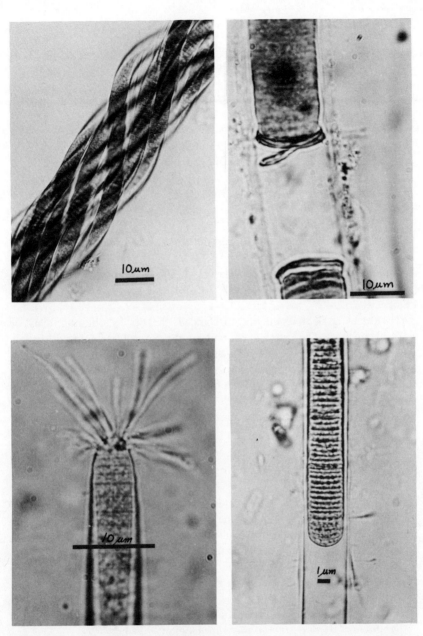

Figs. 11, 12, 13, 14. *Microcoleus lyngbyaceous.* Figure 11 (upper left) shows a fascicle of intertwined trichomes without a distinct sheath that can be keyed out to *Oscillatoria curviceps.* Figure 12 (upper right) shows a single trichome of *Lyngbya majescula* demonstrating the thick sheath and the platelike cells. Figure 13 (lower left) shows *Hydrocoleum lyngbyaceum,* which is characterized by a tuft of hairs at the tip cell. Figure 14 (lower right) shows *Lyngbya Hieronymusii,* which has a distinct sheath.

Oscillatoria obscura Bruhl and Biswas
Oscillatoria margaritifera Kützing
Oscillatoria subbrevis Schmidle
Oscillatoria limosa Agardh *ex* Gomont
Lyngbya confervoides C. Agardh *ex* Gomont
Lyngbya sordida (Zanardini) Gomont
Lyngbya martensiana Meneghini *ex* Gomont
Lyngbya majuscula (Dillwyn) Harvey
Lyngbya aestuarii Liebmon *ex* Gomont
Lyngbya cinerescens Kützing
Lyngbya Hieronymusii Lemmermann
Lyngbya semiplena (C. Agardh) J. Agardh
Hydrocoleum lyngbyaceum Kützing *ex* Gomont
Figs. 11, 12, 13, 14

Humm (unpublished), pp. 32, 41, 42, 43, 44, 49; Desikachary (1959), pp. 202, 206, 207, 209, 221, 285, 297, 305, 313, 314, 315, 318, 347; Drouet (1968), p. 262.

Trichomes straight or curving to spiraled, the tip cells blunt, conical or short attenuated. The outer membrane of the tip cell becoming thickened. Dense protoplasm and granules aligned along walls of each cell. The trichome, in part, showing constrictions at cross walls, measuring 3.0-80.0 μm in diameter. Sheath material hyaline or becoming pigmented (yellow to brown).

Plants corresponding to the above synonyms were collected at Point of Rocks, Venice Jetty, jetties at Ana Maria Island, and at the mouth of the Homosassa River.

Microcoleus vaginatus (Vaucher) Gomont

Drouet (1968), p. 226.

Trichomes 2.5-9.0 μm in diameter, straight or curved, conspicuously attenuated at the tips. Terminal cell conical with truncated end, the outer membrane thickened into a rigid cup-convex disk. Dense protoplasm with granules found along the cross walls. Cells usually quadrate, 1-10 μm in length.

A common species found in sand samples of open coast beaches (Venice Jetty) and as low tufted growth on rocks.

Oscillatoria erythraea (Ehrenberg) Kützing

Synonyms: *Trichodesmium thiebautii* Gomont
            Skujaella thiebautii (Gomont) J. De Toni
Fig. 15

Humm (unpublished), p. 45; Tilden (1910), p. 84; Desikachary (1959), p. 243; Drouet (1968), p. 212.

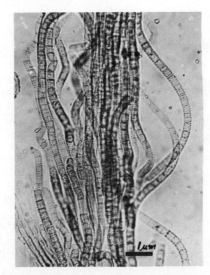

Fig. 15. *Oscillatoria erythraea (Trichodesmium thiebautii)*. The trichomes are entangled to form a fascicle. No distinct gelatinous sheath is visible.

Trichomes 3-30 µm in diameter, cells 2-27 µm in length, the cells over one-third as long as broad, pseudovacuoles common in protoplasm. In typical planktonic forms the trichomes forming ropelike masses.

Common along the entire west coast, especially as planktonic masses during "red tide" outbreaks.

Oscillatoria lutea Agardh

Synonym: *Lyngbya lutea* (Agardh) Gomont

Humm (unpublished), p. 39; Desikachary (1959), p. 310; Drouet (1968), p. 185.

Trichomes 2.5-10 µm in diameter, cells less than one-third as long as broad, trichomes lacking constrictions at cross walls, not uniform in diameter, apex not tapering, sheath hyaline.

Plants forming colonies as crusts on rocks and pneumatophores of mangroves along west coast.

Oscillatoria princeps Vaucher

Synonym: *Oscillatoria proboscidea* Gomont

Desikachary (1959), p. 211; Drouet (1968), p. 174.

Trichomes 8-20 µm in diameter or more, the ends conspicuously tapered, the cells up to one-twelfth as long as broad, usually 3-8 µm long. Cytoplasm homogeneous, walls not granulated. Sheath material hyaline to pigmented (yellow to brown), sheath not distinct.

Collected in sand samples of littoral zone at Siesta Key (Point of Rocks).

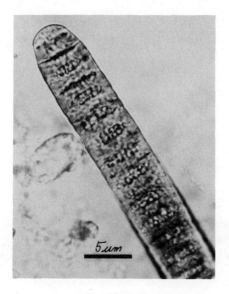

Fig. 16. *Porphyrosiphon Notarisii (Oscillatoria nigro-viridis)*. A tip of a filament showing arrangement of cyanophycean granules along the cross walls.

Porphyrosiphon Notarisii (Meneghini) Kützing
 Synonyms: *Oscillatoria nigro-viridis* Thwaites *ex* Gomont
 Oscillatoria annae van Goor
 Phormidium truncicola Ghose
 Fig. 16
 Humm (unpublished), p. 33; Desikachary (1959), pp. 202, 203, 258; Drouet (1968), p. 143.

Trichomes straight or curving, the ends becoming attenuated through the tapering of two or more cells, the end cell rotund to sharply conical. Trichomes 3-40 μm in diameter, cells about one-third as long as broad, appearing as flat disks. Sheath hyaline to brightly pigmented (brown, red, blue).

The plants comprising a mass of filaments, common in brackish water on mud or limestone outcroppings along the mouths of rivers.

Schizothrix arenaria (Berkeley) Gomont
 Synonyms: *Oscillatoria laetevirens* Crouan *ex* Gomont
 Microcoleus chthonoplastes (Mertens) Zanardini *ex* Gomont
 Microcoleus tenerrimus Gomont
 Phormidium abronema Skuja
 Symploca cartilaginea (Montagne) Gomont
 Symploca laete-viridis Gomont
 Figs. 17, 18
 Humm (unpublished), pp. 31, 53; Tilden (1910), p. 155; Desikachary (1959), pp. 213, 257, 335, 339; Drouet (1968), p. 109.

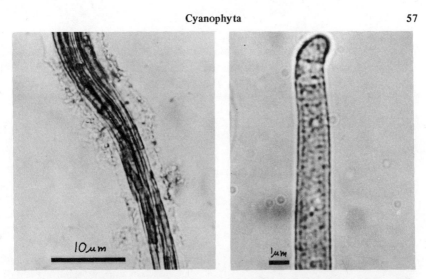

Fig. 17. (left) *Schizothrix arenaria (Microcoleus chthonoplastes)*. The trichomes are grouped together into a fascicle with a distinct, common sheath. Fig. 18. (right) *Schizothrix arenaria*, showing the bent, thick-walled tip cell of the trichome, can be keyed out to *Oscillatoria laetevirens*.

Trichomes varying in diameter between 1-6 μm, constricted at the cross walls, with tip cells tapering and abruptly conical. Cells longer than broad to quadrate, 2-10 μm long. Tip cells conical, the outer wall thin, one to many trichomes in hyaline laminate mucus sheath.

Forming pads or coating over rocks and in tide pools at Gull Rock, mouth of the Homosassa River.

Schizothrix calcicola (Agardh) Gomont
    Synonyms: *Oscillatoria jasorvensis Vouk*
               Oscillatoria foreaui Frémy
               Oscillatoria amphibia Agardh *ex* Gomont
               Lyngbya epiphytica Hieronymus
               Lyngbya nordgaardhii Wille
               Lyngbya holdenii Forti
               Lyngbya digueti Gomont
               Phormidium crosbyanum Tilden
               Phormidium fragile (Meneghini) Gomont
    Fig. 19
    Humm (unpublished), pp. 30, 34, 35, 38; Tilden (1910), p. 96; Desikachary (1959), pp. 219, 221, 229, 253, 284, 286, 287, 310; Drouet (1968), p. 27.

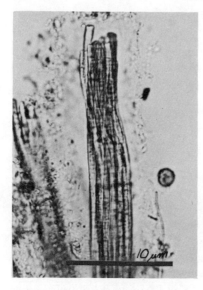

Fig. 19. *Schizothrix calcicola*. A trichome of elongated cells usually found in a common sheath and identified as *Phormidium crosbyanum*.

Trichomes varying in pigmentation, sometimes constricted at nodes (cross walls), 0.2-3.5 µm in diameter, straight or spiraling. Cells quadrate (cubed) or shorter, sometimes longer than broad, 0.2-6 µm long. Protoplasm homogenous to granulose. Terminal (apical) cells rotund to swollen, not tapering, the outer wall thin. Sheath present (visible) or absent.

Specimens keyed out to the synonyms were found along the coast in various littoral sites, on rocks and pneumatophores and bases of salt-marsh grasses. One form (*Phormidium crosbyanum*) typically forms thick gelatinous pads almost resembling a tunicate growing on rocks in the littoral zone.

Schizothrix mexicana Gomont

Synonym: *Lyngbya gracilis* Meneghini *ex* Gomont

Humm (unpublished), p. 40; Desikachary (1959), p. 285; Drouet (1968), p. 87.

Trichomes 4-60 µm in diameter (specimens corresponding to the synonym *L. gracilis* about 8 µm in diameter), sometimes constricted at cross walls, cells 2-10 µm long, terminal cell with thin outer wall. Protoplasm dense, with no regular distribution of granules along walls. Sheath material hyaline.

Common epiphyte on pneumatophores, other algae in Sarasota Pass (east side of Ana Maria Island), and on rocks (Point of Rocks).

Spirulina subsalsa Oersted

Synonyms: *Spirulina major* Kützing
          Spirulina subtilissima Kützing

Spirulina labyrinthiformis (Gmelin) Gomont
Spirulina princeps W. and G. S. West

Fig. 20

Humm (unpublished), p. 28; Tilden (1910), pp. 87, 88; Desikachary (1959), pp. 193, 195, 196, 197; Drouet (1968), p. 16.

Plants a single cell in the form of a trichome, without sheath, straight or more commonly coiled or spiraled. Cells 0.4-4.0 μm in diameter, up to 300 μm in length. Coiling variable, from loose to tight spirals.

Specimens common in littoral sand and rocks along the entire west coast.

Family Nostocaceae

Anabaena fertilissima Rao

Fig. 21

Desikachary (1959), p. 398.

Plants consisting of solitary trichomes, the cells 4-8 μm in diameter, compressed at sites of cellular attachment. Heterocyst quadrate, single with no adjacent akinetes, 6 μm in diameter. Akinete when present terminal, to 10 μm in diameter, slightly ovoid in shape.

Collected in 0.5 m water in a grass bed, Sarasota Pass, Ana Maria Island.

Hormothamnion enteromorphoides Grunow *ex* Bornet *et* Flahault

Humm (unpublished), p. 55; Tilden (1910), p. 205; Desikachary (1959), p. 433.

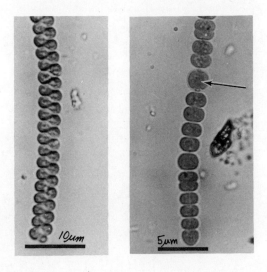

Fig. 20. (left) *Spirulina subsalsa*, a tightly coiled trichome consisting of but one cell. Fig 21. (right) *Anabaena fertilissima*. Arrow denotes heterocyst.

Trichomes up to 8 μm in diameter, cells up to 10 μm long, oval. Heterocysts barrel shaped, to 10 μm in diameter and 12 μm long. Terminal cell conical, blunted tip, little constriction between cells, sheaths distinct.

A common plant of sand, in quiet water (Sarasota Pass, Ana Maria Island; Gull Rock, mouth of Homosassa River).

Nodularia spumigena Mertens var. **major** (Kützing) Bornet *et* Flahault
    Tilden (1910), p. 185; Desikachary (1959), p. 424.

Trichomes 12-20 μm broad, straight or angled, forming entangled masses, regular in diameter and not constricted. Cells to 12 μm in diameter, one-third as long as broad, apical cells and cells adjacent to heterocysts conical to pointed. Heterocysts intercalary, compressed to 14 μm in diameter and 8 μm in length.

Forming entangled masses among mangroves at Sarasota Pass, Ana Maria Island. *Nodularia harveyana* (Thwaites) Thuret has been recorded once from Point of Rocks, Florida (Eiseman, in preparation).

Family Rivulariaceae

Calothrix confervicola (Roth) C. Agardh
    Humm (unpublished), p. 60; Tilden (1910), p. 256.

Trichomes clumped together forming fascicles to 3 mm in height, with basal heterocysts only, and trichome tapering to a fine hair. Basal (lower) cells bulbous, 12-20 μm in diameter, one-third to one-fourth as wide as long; false branching not common.

Frequently found as fasciculated tufts on larger algae (especially *Sargassum*), on shells and rocks in such estuarian regions as the mouth of the Homosassa River.

Calothrix crustacea Thuret
    Figs. 22, 23
    Humm (unpublished), p. 62; Tilden (1910), p. 264; Desikachary (1959), p. 523.

Dark green plants consisting of trichomes to 30 μm in diameter, little constricted at cross walls, showing false branching. The trichomes attenuated at apices, sometimes forming long hyaline hairs. Cells one-third to one-half as long as broad; heterocysts basal or intercalary, at site of false branching the heterocysts one-half as long to spherical in shape. Sheath distinct, hyaline, up to 5 μm thick, spreading open near tapering portion of trichome and lamellated (sometimes) at base.

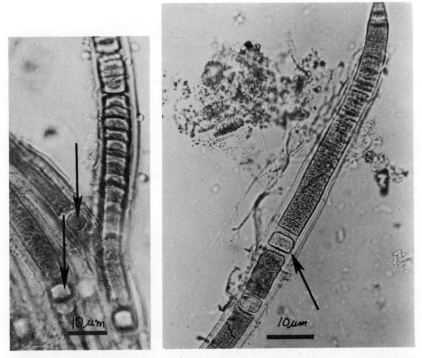

Figs. 22, 23. *Calothrix crustacea*. The basal portion of the filaments (fig. 22, left) shows a number of intercalary heterocysts (arrows) whereas the upper portion (fig. 23, right) shows a heterocyst (arrow) and the abrupt tapering of the trichome.

A common turf forming blue-greens of the littoral zone, found on rocks (Point of Rocks, Venice Jetty) and on shells.

Calothrix parietina (Naegeli) Thuret
Fig. 24
Humm (unpublished), p. 60; Tilden (1910), p. 269; Desikachary (1959), p. 538.

Trichomes 10-16 μm in diameter at base, constricted at nodes; usually tapering to a hair; heterocysts basal or intercalary. Cells isodiametric (quadrate) in central region of trichome, 6-10 μm in diameter; sheath not thickened but distinct, usually yellow to brown in color.

Common in sand collections of littoral regions along open coast (Siesta Key).

Calothrix pilosa Harvey
Humm (unpublished), p. 61; Tilden (1910), p. 263.

Trichome tapering to a hair, but not attenuated, rather abrupt and therefore not obvious. Trichome large, 15-40 μm in diameter, with constrictions appearing at cross walls in older (larger) filaments. Heterocysts basal and intercalary, false branching common.

Occasional at Gull Rock at the mouth of Homosassa River.

Dichothrix fucicola (Kützing) Bornet *et* Flahault
Humm (unpublished), p. 64; Tilden (1910), p. 279.

Trichomes in dense tufts; cells 6-12 μm in diameter near heterocysts; heterocysts basal and intercalary, to 30 μm in diameter, spherical at base and elongate-quadrate in intercalary positions, Sheath thick throughout the filament, brown to yellow in color.

Found as epiphytes on Sargassum (mouth of Homosassa River) or on rocks (Venice Jetty).

Rivularia atra Roth
Humm (unpublished), p. 66; Tilden (1910), p. 289.

Colonies of filaments spherical and solid, not becoming hollow with aging;

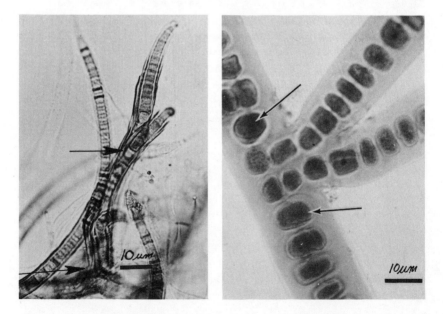

Fig. 24. (left) *Calothrix parietina*. False branching (arrows) is evident in the tapering filaments. Fig. 25. (right) *Stigonema aerugineum*. True branching and "pit connections" are evident. Arrows denote heterocysts present on either side of true branches.

trichomes forming a tuft in the matrix; heterocysts basal. Trichomes 2-5 μm in diameter and tapering to a hair.

Collected on pneumatophores of black mangroves, Sarasota Pass, Ana Maria Island.

Rivularia nitida Agardh
Humm (unpublished), p. 67; Tilden (1910), p. 287.

Colonies becoming spherical, expanded, soft and hollow. Trichomes 2-5 μm in diameter, terminating in a long thin hair. Sheath thickened at base, having a yellow to brown color. Heterocysts basal.

Collected as small, rugose, bright green buttons on Gull Rock, mouth of Homosassa River.

Rivularia polyotis (Agardh) Bornet *et* Flahault
Humm (unpublished), p. 66; Tilden (1910), p. 286.

Colonies hollow and pulvinate with age. Trichomes usually 8-15 μm in diameter in the middle, then tapering to a hair; constrictions present at cross walls, heterocysts to 15 μm in diameter, slightly spherical.

On pneumatophores of black mangroves at Sarasota Pass, Ana Maria Island.

Family Stigonemataceae

Mastigocoleus testarum Lagerheim
Humm (unpublished), p. 58; Tilden (1910), p. 237; Desikachary (1959), p. 575.

Plants penetrating limestone and shells; consisting of trichomes with true branching, the branches tapering to a hair or having terminal heterocysts; the branches slightly smaller in diameter than the main axis (3-6 μm versus 6-10 μm in diameter), heterocysts usually 6-18 μm in diameter.

Occasional on shell fragments at Point of Rocks, Siesta Key.

Stigonema aerugineum Tilden
Fig. 25
Tilden (1910), p. 245; Desikachary (1959), p. 606.

Trichomes to two rows of cells in linear arrangement in a gelatinous, hyaline sheath. The cells discoid, from 15 μm in diameter and 6 μm in length; true branching common; uniseriate branches formed at right angles to main axis. Heterocysts intercalary, to 8 μm in diameter.

Collected once in a tide pool organically rich from bird droppings at Gull Rock, mouth of the Homosassa River.

5

Chlorophyta and
Chrysophyta

SPECIES DESCRIPTIONS

Taxonomic literature and other drawings of green algal species listed here are found in Humm and Taylor (1961) and W. R. Taylor (1960).

ORDER TETRASPORALES
Family Palmellaceae

Pseudotetraspora antillarum Howe
  Taylor (1960), p. 45; Humm and Taylor (1961), p. 342.
  A small, soft, globular or flattened gelatinous mass, usually 2-20 mm in diameter, yellowish brown in color. Cells globose or ellipsoid, mostly 3-7 μm in diameter, spaced in gelatinous material, sometimes in patterns of four.
  Attached to other algae, sea grasses, or hard substrata.

ORDER ULOTRICLIALES
Family Gomontiaceae

Gomontia polyrhiza (Lagerheim) Bornet *et* Flahault
  Taylor (1960), p. 53; Humm and Taylor (1961), p. 342.

Plants filamentous, branched, cells 4-8 μm in diameter; sporangia produced at the surface, 30-40 μm in diameter.

Littoral to sublittoral, usually penetrating shells or limestone.

Family Chaetopeltidaceae

Diplochaete solitaria Collins

Taylor (1960), p. 53.

Plant consisting of single cell or clusters of cells in gelatinous sheaths, the cells up to 35 μm in diameter and with distinct setae (usually two) arising from base or side of cell and tapering.

Collected as an epiphyte of *Thalassia* blades off Homosassa River.

Family Chaetophoraceae

Entocladia flustrae (Reinke) Batters

Humm and Taylor (1961), p. 338.

Microscopic, consisting of irregular and densely branching filaments which ultimately form a central membrane with filamentous margin. Typically on the surface of Bryozoa. Plastid parietal with one pyrenoid. Central cells irregularly polygonal, usually 7-12 μm in diameter, occasionally larger.

Known from northern Gulf of Mexico and probably present in Tampa Bay area.

Entocladia viridis Reinke

Taylor (1960), p. 60; Humm and Taylor (1961), p. 338.

Consisting of microscopic, irregular, branched filaments, 3-8 μm in diameter; filaments creeping upon the surface of the host plant or within the gelatinous coating of the host. Cells one to six diameters long, usually irregularly swollen and contorted, with a single pyrenoid.

Occurs on or in a wide variety of algae, also in the tunic of Ascidians.

Phaeophila dendroides (Crouan) Batters

Taylor (1960), p. 48; Humm and Taylor (1961), p. 341.

Microscopic, branched filaments, growing horizontally. Upright, colorless, zigzag or slightly spiral hairs produced by some of the cells. Cells 9-40 μm in diameter, 15-50 μm long.

A common epiphyte or endophyte on larger algae, sea grasses, or on shells.

Pringsheimiella scutata (Reinke) Schmidt *et* Petrak
    Taylor (1960), p. 51; Humm and Taylor (1961), p. 339.
    Plant forming a monostromatic disk up to 2 mm in diameter, epiphytic;
cells about 12 μm in diameter but variable in size and shape; cells usually
columnar in center and elongate at edges of plant.
    Collected as an epiphyte on *Thalassia* in grass beds off the Homosassa
River.

Protoderma marinum Reinke
    Taylor (1960), p. 52; Humm and Taylor (1961), as *P. marina,* p. 341.
    Plant a minute disk; forming extensive coatings on rocks; firmly attached.
Cells 5-12 μm in diameter, polygonal by mutual pressure except at the margins where short, radiating filaments may be produced.
    In rocky areas forming a green band at the high tide line and in the splash
zone.

Ulvella lens Crouan
    Taylor (1960), p. 52; Humm and Taylor (1961), p. 341.
    Plant forming a radiating disk 0.1-1.0 mm in diameter; closely appressed,
radiating filaments mostly 2-5 (to 6) μm in diameter; cells elongate-rectangu-
lar in shape (occasionally square), 5-12 (to 20) μm long. The marginal cells
sometimes forked. Disk monostromatic in the radiating portion (edge), but
several cells thick at the center where the cells are irregular in arrangement.
Seen from above, the plant more or less rounded. Plastid a parietal disk filling
the entire face of the cell.
    Plants epiphytic on larger algae and sea grasses and epizoic on ascidians
and barnacles. Also found on stones and shells in 1-3 m of water.

Family Ulotrichaceae

Stichococcus marinus (Wille) Hazen
    Humm and Taylor (1961), p. 335.
    Filaments unbranched, 5-8 μm diameter and lacking a modified basal cell.
Cells one to two and one-half diameters long with a parietal chromatophore
covering about half the cell face and having one pyrenoid. Asexual reproduction by biflagellate zoospores having a single eyespot (stigma).
    Collected attached to shells and limestone in the littoral zone, especially in
brackish water. Collected on inner side of jetty forming the Venice break-
water.

ORDER ULVALES

Family Ulvaceae

Enteromorpha chaetomorphoides Børgesen

Taylor (1960) p. 57.

Plants appearing filamentous, terete, simple or proliferously branched. Tips of branches uniseriate and basal regions consisting of two to three cells in cross section about 45 μm in diameter. Cells 16 μm in diameter, arranged in longitudinal rows, and rectangular in shape.

Found entangled with other algae or attached on roots of mangroves or shells in brackish water, usually in sheltered littoral zones.

Enteromorpha clathrata (Roth) J. Agardh

Taylor (1960) p. 58; Humm and Taylor (1961) p. 343.

Plants light green, to about 4 cm long, slender axes, repeatedly branched; branches tapering from base to apex but not producing a uniseriate tip. Cells in distinct longitudinal rows and rectangular in shape; 18-28 μm in diameter and 13-38 μm long. The plastid distinctly smaller than the cell face.

Characteristic of the winter flora along the west coast of Florida and occasionally collected in the brackish water of Tampa Bay and near the mouth of the Homosassa River.

Enteromorpha compressa (Linnaeus) Greville

Taylor (1960), p. 60.

Plant usually gregarious; fronds to 30 cm long, tubular, compressed or collapsed; upper mature portions 2-20 mm wide, tapering below with several branches from basal region; cells not in longitudinal rows, 10-15 μm in diameter, walls thin.

In Tampa Bay present during the winter; attached to rocks and shells.

Enteromorpha flexuosa (Wulfen) J. Agardh

Taylor (1960), p. 61; Humm and Taylor (1961), p. 345.

Plants unbranched (rarely branched at the base); fronds inflated or flattened; the cells with thick walls and usually in longitudinal rows, 10-28 μm in diameter.

Continuous along the entire Florida west coast. A plant of the littoral and shallow waters found attached to rocks and shells.

Enteromorpha intestinalis (Linnaeus) Link

Fig. 26

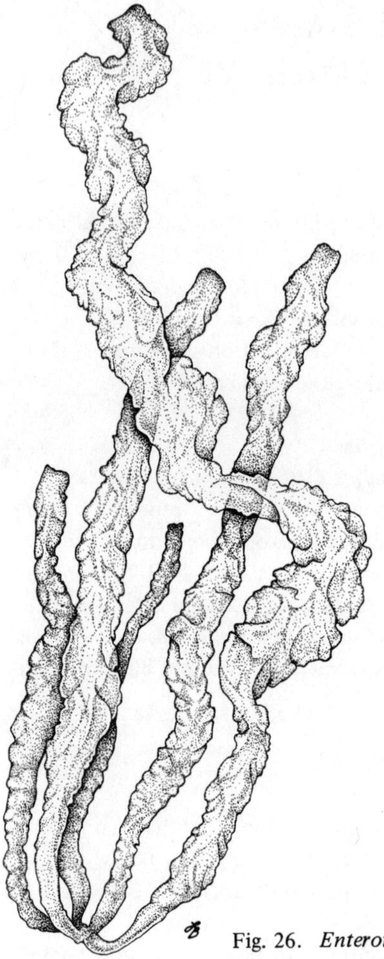

Fig. 26. *Enteromorpha intestinalis.   x 1.5*

Taylor (1960), p. 62; Humm and Taylor (1961), p. 345.

Similar to *E. flexuosa* in gross aspect, except that the plants are usually unbranched, larger, and more contorted and constricted. Cell walls thin, cells not in longitudinal rows and scattered.

A cold-water species present during the winter months.

Enteromorpha lingulata J. Agardh

Taylor (1960), p. 60; Humm and Taylor (1961), p. 347.

Plant cylindrical, tufted or turf-forming, 3-7 cm tall, sparingly branched at the base, rarely branched in the upper parts. Branches progressively larger from the base upward; cells rectangular and in longitudinal rows. Apparently present throughout the year; attached to hard substrate.

Enteromorpha linza (Linnaeus) J. Agardh
Taylor (1960), p. 63; Humm and Taylor (1961), p. 347.

Plants unbranched; the short hollow stipe forming a flat blade; the blade tapering, lanceolate or linear-lanceolate in shape; plants usually to 5 cm tall. The blade portion is like *Ulva*, being two cells thick except at the edges where the two layers separate.

Collected on oyster shells near Homosassa River in winter months.

Enteromorpha prolifera (Müller) J. Agardh
Taylor (1960), p. 59; Humm and Taylor (1961), p. 348.

Plants sparingly branched throughout, occasionally rebranched; cells 10-15 μm in diameter, 15-18 μm long, and in longitudinal series in younger parts.

Attached to shells and rocks in littoral or shallow sublittoral regions in winter and spring.

Enteromorpha ramulosa (J. E. Smith) Hooker
Taylor (1960), p. 60; Humm and Taylor (1961), p. 348.

Plant tubular, 10-30 cm tall, to 1 mm in diameter, rather stiff; numerous branches bearing short, spinelike branchlets. Cells rounded, 15 μm in diameter, and almost completely filled by the plastid, only arranged in longitudinal rows in the ultimate divisions.

Present in the winter; attached to hard substrate.

Enteromorpha salina Kützing, var. **polyclados** Kützing
Taylor (1960), p. 56; Humm and Taylor (1961), p. 348.

Plant entangled, slender, sparingly branched throughout or only at the base; the lower portions with short, tapering branches, uniseriate at the tips. Cells cuboidal or shorter than broad, about 16 μm wide, in longitudinal series throughout. Cell wall thickened on both sides.

Known in the northern Gulf; probably present in the Tampa Bay area during the winter months.

Monostroma oxyspermum (Kützing) Doty
Taylor (1960), p. 64; Humm and Taylor (1961), p. 352.

Plant a thin monostromatic sheet, 20-25 μm thick, broadly lanceolate to orbicular; in surface view the cells scattered, polygonal or rounded, 8-17 μm in diameter; cells round to oval in cross section. Until recently plants of this species from Florida have been referred to as *M. latissimum* (Kützing) Witt-rock. Taylor (1957) discusses this problem.

Attached to mangrove roots, rocks and other substrate, especially in brack-ish water.

Ulva fasciata Delile

Taylor (1960), p. 66; Humm and Taylor (1961), p. 351.

A distromatic sheet, typically divided into long, linear segments; cells taller than wide in section. Blade thicker in the center, where the cells are tallest, and thinner toward the margins, where the cells may be nearly as wide as tall.

Typically found on rock jetties exposed to wave action. Known from a variety of sites.

Ulva lactuca Linnaeus

Fig. 27

Taylor (1960), p. 65; Humm and Taylor (1961), p. 351.

A distromatic sheet, orbicular to irregular in outline, not divided into linear segments; sheet about the same thickness throughout, attached at first, later drifting.

Var. **latissima** (Linnaeus) De Candolle is a thin blade with cells that are isodiametric or sometimes wider than tall. The plastids are peripheral in the cells. This variety is more characteristic of salt marshes or bays with muddy waters and bottoms.

Var. **rigida** (C. Agardh) Le Jolis is distinguished by having firm, thick (60-110 μm) blades. The cells are taller than wide and usually lanceolate. Plastids usually cover the surface of the cells. Old plants may become perfor-ated.

Both varieties are common in the Tampa Bay area, reaching their peak of development in the spring; they are especially conspicuous attached to vari-ous substrate in the littoral zone.

ORDER SIPHONALES

Family Derbesiaceae

Derbesia vaucheriaeformis (Harvey) J. Agardh

Taylor (1960), p. 128; Humm and Taylor (1961), p. 359.

Plants forming dense tufts, filaments dichotomously branched, 1-5 cm tall, 35-50 μm in diameter; apices obtuse. Sporangia borne on the lower parts of the plant, apparently replacing a branch. Sporangia ovate to pyriform, 75-115

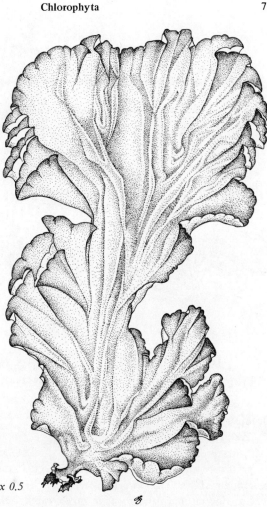

Fig. 27. *Ulva lactuca.* x 0.5

μm by 130-180 μm (in Florida material), with a short stalk cell about 15 μm in diameter and two diameters long; sporangia contain about 64 ovoid spores; the spores are 16-18 μm in diameter.

Forming mats over mud and sand on mangrove (*Rhizophora*) islands of Tampa Bay, apparently contributing to the stability of substrate (Dawes, 1965). Also growing in deeper, offshore waters. According to Sears and Wilce (1970), this species may be a form of *D. marina.*

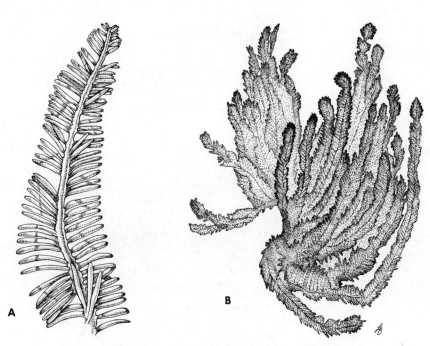

Fig. 28. *Bryopsis pennata.* A, *x 5* B, *x 1.5*

Family Bryopsidaceae

Bryopsis hypnoides Lamouroux
Taylor (1960), p. 130.

Plants erect to 10 cm; dark green; the branches of same diameter or progressively smaller than main axes; much branched, branches irregularly placed, about 40-80 μm in diameter.

Growing on mangrove roots in quiet water, but also found in open, heavy surf on rocks (e.g., Mullet Key, Fort DeSoto Park).

Bryopsis pennata Lamouroux
Fig. 28
Taylor (1960), p. 132; Humm and Taylor (1961), p. 360.

Fronds coenocytic, featherlike in appearance, developing from a rhizoidal base, 2-5 cm tall; branching distichous; determinate branchlets of uniform length, their bases constricted.

Common on rock jetties and breakwaters where wave action is pronounced. Older material may be keyed out to *B. ramulosa* and *B. duchassaignii* (Eiseman, in preparation).

Bryopsis plumosa (Hudson) C. Agardh
    Taylor (1960), p. 131.

Plants erect to 10 cm tall, usually olive green, fronds 1 cm in width; the axes only branched toward the tip and becoming progressively shorter; branching distichous; fronds spatulate, triangular, or pyramidal in shape; branches 60-100 μm in diameter, sharply constricted at the base.

Found on sublittoral rocks exposed to moderate surf action (e.g., Mullet Key).

Family Caulerpaceae

Caulerpa ashmeadii Harvey
    Fig. 29
    Taylor (1960), p. 142; Humm and Taylor (1961), p. 360.

Fig. 29. *Caulerpa ashmeadii.* x 0.5

Erect branches distichous, mostly 4-13 cm tall, producing two rows of cylindrical or slender, clavate branchlets; branchlets 10-18 mm long, with blunt or rounded tips. The erect branches are quite variable, some have few branchlets, others have secund branchlets and, occasionally, there are no branchlets.

A common but usually inconspicuous alga among sea grasses in shallow water.

Caulerpa cupressoides (West) C. Agardh

Taylor (1960), p. 146; Humm and Taylor (1961), p. 362.

Erect axes dichotomously branched and bearing short branchlets in two, three, or more ranks; the branchlets are conical, flattened, or cylindrical in shape, and they are always sharply pointed. The species is highly variable. Taylor (1960) lists all varieties and two forms for Florida.

Var. **flabellata** Børgesen has branchlets one to three times as long as the midrib. All specimens from Tampa Bay northward seem to fit this variety.

Occurs among sea grasses and on limestone in shallow water.

Caulerpa fastigiata Montagne

Taylor (1960), p. 136.

Plants forming clumps or mats; erect portions not differentiated from rhizome; filaments about 150 μm in diameter and dichotomously forked to irregular in branching.

Plants may be mistaken for *Bryopsis hypnoides* until the wall trabeculae are viewed with the light microscope. Collected on coquina outcroppings in the littoral zone at Siesta Key and also on mangrove roots.

Caulerpa lanuginosa J. Agardh

Taylor (1960), p. 145.

Erect axes densely covered with short branchlets; axes about 4-8 cm tall; the rhizome is also covered with short branchlets and is difficult to remove from the substrate. Branchlets randomly arranged around the axes, 2-4 mm long, and pointed at the apex.

A single collection was made offshore at Homosassa River; growing in sand among blades of *Thalassia.*

Caulerpa mexicana (Sonder) J. Agardh

Fig. 30

Taylor (1960), p. 141; Humm and Taylor (1961), as *C. crassifolia,* p. 362.

Erect branches pinnate, mostly 2-10 cm tall; branchlets flattened, 1-2 cm wide, often curved, tapering to a point, may be sickle shaped.

On mangrove roots, rocks, or muddy sands.

Fig. 30. *Caulerpa mexicana.* x 1.5

Caulerpa paspaloides (Bory) Greville
   Taylor (1960), p. 149; Humm and Taylor (1961), p. 363.
   Rhizomes bearing palmately or dichotomously branched erect axes; main axes naked to the point of branching. Branches covered with branchlets in three to four alternating ranks, these in turn with one or two ranks of pinnules along the upper side; the pinnules are usually sharply pointed. The species is extremely variable in morphology and size; plants range in height from 1-15 cm.
   Var. **minima** Humm and Taylor is about 1 cm tall, forming an entangled mass on rocks or sponges. Although reported only in the northern Gulf, it probably occurs in the Tampa Bay area.
   On limestone, rock, or mud bottom, commonly among sea grasses.
   Var. **wurdemanni** Weber-van Bosse has axes about 8 cm high with branches to 10 cm long; branchlets two-ranked, subopposite, or scattered. The branchlets have two ranks of unilateral, simple pinnules.
   Common offshore in grass beds.

Caulerpa peltata Lamouroux
   Taylor (1960), p. 155; Humm and Taylor (1961), p. 364.

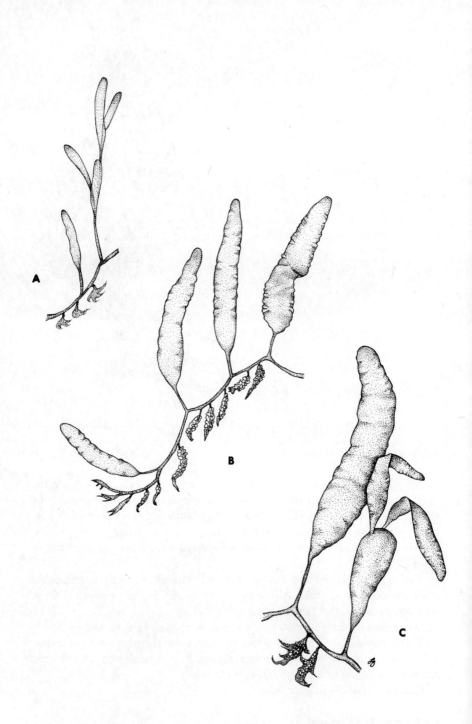

Fig. 31. *Caulerpa prolifera*. Three forms are shown: A, f. *zosterifolia, x 0.5*; B, the typical form, *x 0.5*; and C, f. *obovata, x 0.5.*

Erect branches 5-15 mm tall, simple or branched; the branchlets peltate, 3-8 mm in diameter, and on short pedicels.

On rocks in shallow water, St. Petersburg Beach. This species is considered a variety of *C. racemosa* by Eubank (1946, p. 421).

Caulerpa prolifera (Forsskål) Lamouroux
Fig. 31

Taylor (1960), p. 140; Humm and Taylor (1961), p. 364.

Upright branches in the form of flat blades; blades linear to lanceolate, sometimes proliferous, 4-15 cm tall, 7-20 mm wide or more.

F. **obovata** J. Agardh. A coarse form of the species with dark green blades; blades 15-25 mm broad, to 100 mm long, and proliferous.

F. **zosterifolia** Børgesen. Plants with lanceolate blades less than 6 mm wide; blades arise from a small rhizome (1-3 mm in diameter).

All three forms, including the typical form, are found in grass beds. A deepwater form, described as *C. ollivieri* Dostal by Hine and Humm (1971), has been reported from the west coast; however, *C. ollivieri* appears to be a diminutive form of f. *zosterifolia* (Barilotti, 1970) and therefore is not listed here as a separate species.

Caulerpa racemosa (Forsskål) J. Agardh
Fig. 32

Taylor (1960), p. 151; Humm and Taylor (1961), p. 365.

Erect axes with numerous branchlets; the ends flattened, swollen, or spherical. A highly variable species with seven varieties reported for Florida (Taylor, 1960).

Var. **clavifera** (Turner) Weber-van Bosse. Erect axes 1-3 cm tall, branchlets 0.5-1.0 mm long, gradually expanded to a rounded end. Collected adrift at Clearwater Beach, Florida.

Var. **laetevirens** (Montagne) Weber-van Bosse. A large plant (12-30 cm tall), branchlets widely spaced; branchlets are subcylindrical and have a rounded apex. The plant is common on rocky substrate, particularly at Point of Rocks (Eiseman, in preparation).

Var. **uvifera** (Turner) Weber-van Bosse. Erect axes 1.5-2.5 cm tall, branchlets crowded, swollen at tips to form a rounded top, and covering the axes. Collected at Siesta Key.

Caulerpa sertularioides (Gmelin) Howe

Taylor (1960), p. 144; Humm and Taylor (1961), p. 365.

Erect axes usually with two rows of distichous branchlets producing a flat blade; branchlets may occur in three or more ranks and have pointed tips.

F. **brevipes** (J. Agardh) Svedelius. Erect axis rarely forked, the blade sessile or nearly so.

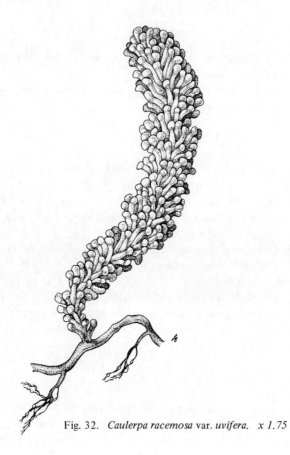

Fig. 32. *Caulerpa racemosa* var. *uvifera.* x 1.75

Occurs on natural limestone or jetty rocks; a very elongate form is found in depths of 30-40 m.

Family Codiaceae

Avrainvillea asarifolia Bϕrgesen
  Taylor (1960), p. 161.
  Blades dull green, smooth, and thin; plant to 20 cm tall, stipe 5-10 cm, blade 8-10 cm long by 15 cm broad; blade orbicular to oblong; filaments smooth, in medulla 20-30 μm in diameter, in cortical region 8-13 μm in diameter and moniliform.
  Collected in soft sand flats off Indian Bluff Island, Tarpon Springs.

Avrainvillea elliottii A. *et* E. S. Gepp
Taylor (1960), p. 162.

Blades deep green to brown in color; plants to 12 cm in height, although usually smaller; plants have a short, thick stipe 1 cm long and a thin, firm blade to 7 cm wide; filaments 20-30 μm and uniform in diameter to slightly moniliform.

One collection of a number of plants was made in 30 m of water off Sarasota.

Avrainvillea levis Howe
Taylor (1960), p. 162; Humm and Taylor (1961), p. 365.

Stalk 3-5 cm long, 5-6 mm in diameter; blade flat, 4-7 cm tall, about 5-9 cm wide, soft, spongy in texture, zonate, and broadly rounded to reniform in shape, often with a slightly cordate base. Plant dull green in color, composed of dichotomously branched filaments that are constricted above the forks. Surface filaments of the blade rather closely interwoven to form a cortexlike layer; filaments not moniliform to only slightly so in the ultimate divisions. Cortical filaments 10-24 μm in diameter; axial filaments coarser, 20-35 μm in diameter.

Common in the northern Gulf, around St. Theresa, Franklin County. Grows on limestone rock among sea grasses.

Avrainvillea nigricans Decaisne
Fig. 33
Taylor (1960), p. 160.

Plants to 15 cm tall, blackish green, stalked or sessile; blade thick and hairy, irregularly lobed, to 6 cm long; medulla composed of moniliform filaments 5-70 μm in diameter, the corticating filaments of the surface about 30 μm.

Collected from shallow grass flats in upper Tampa Bay.

Boodleopsis pusilla (Collins) Taylor, Joly, *et* Bernatowicz
Taylor (1960), p. 157.

Filaments forming a mat over mud, the filaments dichotomously branched and strongly constricted at the base of each branchlet. The filaments to 90 μm in diameter, septations reported but not found; filament tips about 45 μm in diameter or less. Subspherical sporangialike structure present, about 100 μm in diameter.

Plant forming a thick mat over the mud of red mangrove swamps in association with *Cladophora repens* (Dawes, 1965). The association appears to have a stabilizing effect in the littoral zone.

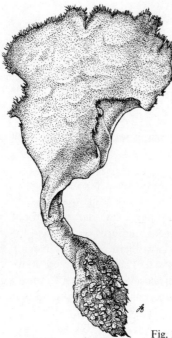

Fig. 33. *Avrainvillea nigricans.* x 1

Codium decorticatum (Woodward) Howe

Taylor (1960), p. 188; Humm and Taylor (1961), p. 367.

Plants 12 cm to 1 m tall; branching dichotomous, the branches typically flattened below the dichotomies. Utricles 300-570 µm in diameter, the end walls not thickened.

Reported from Tampa Bay and vicinity, growing on rocks in shallow, sublittoral waters.

Codium isthmocladum Vickers

Taylor (1960), p. 186; Humm and Taylor (1961), p. 367.

Plants similar to *C. decorticatum,* but branches are terete in cross section below the dichotomies and have utricles 120-460 µm in diameter or more. Utricles usually 250-300 µm. End walls of utricles thickened, 18-56 µm.

A common species, attached to limestone in depths of 10-30 m.

Codium taylori Silva

Fig. 34

Taylor (1960), p. 188; Humm and Taylor (1961), p. 368.

Branches cylindrical, though distinctly flattened below the dichotomies; plants dark green, stiff, densely branched. Utricles spindle shaped to clavate or cylindrical; the outer face of the utricles broadly rounded, 125-300 μm in diameter, the utricles two to five diameters long.

On rocks at Mullet Key, Tampa Bay, and on rocks at the end of the peninsula, Pass-a-Grille, Pinellas County.

Halimeda discoidea Decaisne

Taylor (1960), p. 179; Humm and Taylor (1961), p. 369.

Plants 6-15 cm tall, branched; anchored by a dense cluster of rhizoids. Axis composed of flattened, smooth, orbicular or suborbicular segments, 1-3 cm in diameter. Segments lightly calcified and with entire margins. Central filaments producing expanded utricles at the periphery, which are laterally appressed and polyhedral when viewed from above; utricles 40-60 μm to 85 μm in diameter.

A common deepwater (8-9 m) species usually found in sandy areas.

Halimeda incrassata (Ellis) Lamouroux

Fig. 35

Taylor (1960), p. 181; Humm and Taylor (1961), p. 369.

Differing from *H. discoidea* in that the lower segment of the plant forms an extensive stalk; the plant is branched and composed of flattened to lobed segments. Above the stalk the segments are triangular to ovate or even cylindrical in shape; the segments are 4-10 mm in diameter, three-lobed, entire,

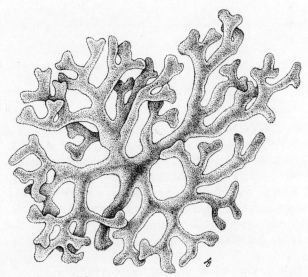

Fig. 34. *Codium taylori.* x 0.5

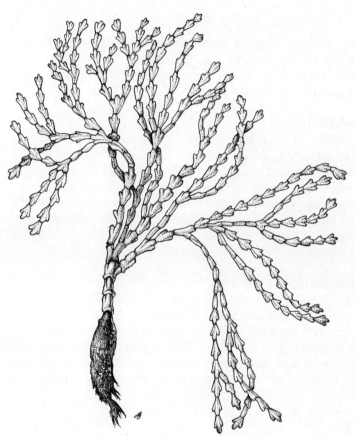

Fig. 35. *Halimeda incrassata.* x 0.5

and usually with prominent ribs or ridges. Utricles 42-84 μm in diameter. Plants 15-25 cm tall.

Common among sea grasses.

Halimeda monile (Ellis *et* Solander) Lamouroux
Taylor (1960), p. 182.

Plants 10-25 cm tall, dark green, firmly calcified, attached in sand by a well-developed rhizoidal system; the lower three segments 1.5-2.0 mm in diameter and may be somewhat flattened. The upper branch segments terete; segments three-lobed where new branches arise; other lobes cylindrical.

An occasional plant growing in offshore grass flats (e.g., Indian Bluff Island and around Anclote Key) in 3-4 m of water.

Halimeda opuntia (Linnaeus) Lamouroux
Taylor (1960), p. 176.

Plants white to green, heavily calcified; growing as massive clumps, young plants with a single holdfast but holdfast later obscured by secondary attachments and random branching. The segments are three-lobed, but not as pronounced as in *H. incrassata;* segments 3-10 mm wide. Medullary filaments fused in pairs at nodes.

A common offshore plant in 12-25 m of water (Dawes and van Breedveld, 1969).

Penicillus capitatus Lamarck
  Fig. 36
  Taylor (1960), p. 171; Humm and Taylor (1961), p. 371.
  Plants 7-15 cm tall, composed of an erect stipe bearing a head of slender, dichotomously branched and lightly calcified filaments. Filaments are densely fasciculate. Filaments of the head 100-200 μm in diameter; the stipe penetrating to the center of the head.

  Common on limestone and among sea grasses in sandy substrates. Deeper plants (3-5 m) often have a long stipe (to 25 cm).

Fig. 36. *Penicillus capitatus.* x 0.75

Penicillus dumetosus (Lamouroux) Blainville
   Taylor (1960), p. 172.
   Plants to 1 dm tall; stalks coarse, to 6 cm in diameter, and often slightly
flattened. Filaments 400-800 μm in diameter, not heavily calcified.
   A common offshore species in 5-10 m.

Penicillus lamourouxii Decaisne
   Taylor (1960), p. 172; Humm and Taylor (1961), p. 371.
   Differing from *P. capitatus* in that the filaments of the head are 350-500
μm in diameter, and more heavily calcified.
   Common on limestone bottoms among sea grasses in 1-3 m of water.

Pseudocodium floridanum Dawes *et* Mathieson
   Fig. 37
   Dawes and Mathieson (1972).
   Plants 3-10 cm in height, bright green in color, soft and flexible; consisting
of a main axis 0.2-0.5 cm in diameter and about 1-4 cm long. Branching
dichotomous, the branches and main axis consisting of 1-10 medullary filaments surrounded by tightly compacted utricles. Medullary filaments moniliform, the swellings 40-70 μm in diameter, the constrictions 14-50 μm in
diameter. Utricles polygonal to hexagonal in surface view, 140-280 μm in
diameter. Gametangialike structures formed from medullary filaments and
not projecting into cortex.
   Dredged in 20-60 m of water during Hourglass cruise (Dawes and van
Breedveld, 1969). The genus is newly recorded for the Northern Hemisphere;
it is also a new species for the genus (Dawes and Mathieson, 1972).

Rhipocephalus phoenix (Ellis *et* Solander) Kützing
   Taylor (1960), p. 174.
   Plants 7-12 cm tall, local specimens only reaching 5 cm; stalks terete,
smooth, and calcified; cap consisting of a tuft of inverted wedge-shaped
blades, to 3 cm long and about 5 mm wide; blades composed of filaments
210-240 μm in diameter at base and 50-100 μm in diameter at the tips.
   Collected from offshore reefs in 3-20 m of water.

Udotea conglutinata (Ellis *et* Solander) Lamouroux
   Fig. 38
   Taylor (1960), p. 165; Humm and Taylor (1961), p. 371.
   Plants 10-15 cm tall; having an erect stipe and a flat, orbicular to fan-
shaped blade. Structurally composed of dichotomous filaments that in the
blade are 24-45 μm in diameter (to 60 μm). Blade calcified; corticating

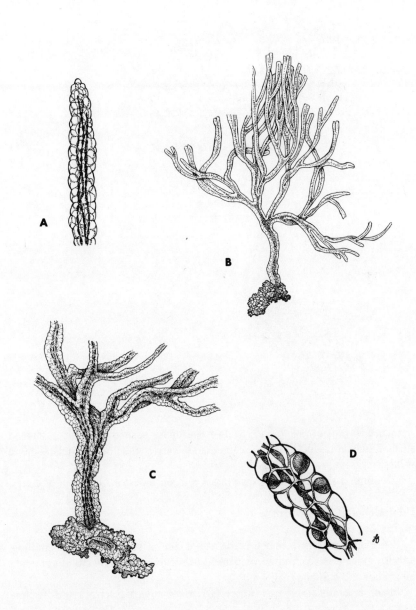

Fig. 37. *Pseudocodium floridanum.* A habit sketch (B, *x 1.5*) shows the general dichotomous branching of the plant. The inner siphonous filaments can be seen in the enlargement of the plant base (C, *x 3.5*) and in a filament tip (A, *x 11.5*). Gametangialike swellings are evident through the noncalcified utricles, as shown by D, *x 35.*

Fig. 38. *Udotea conglutinata.* x 1

branches or projections absent. Stalk has a dense cortex of widely branched surface filaments; the terminal segments are very slender, with thickened end walls.

A common species among sea grasses and on limestone in 2-5 m of water.

Udotea cyathiformis Decaisne
  Taylor (1960), p. 166.
  Plants have a funnel-shaped blade and a central stalk. Blades 3-6 cm long, whitish green, and composed of packed corticating filament, 50 μm in diameter. Stalk 1-2 cm tall.
  Plants collected in deeper water (15 to 30 m) off Sarasota and Anclote Key.

Udotea flabellum (Ellis *et* Solander) Lamouroux
  Taylor (1960), p. 168; Humm and Taylor (1961), p. 372.
  Differing from *U. conglutinata* in that the blade is firm and smooth and not obviously filamentous. Filaments of the blade produce a cortex of densely branched, surface filaments; filaments have truncate and thickened

end walls. Blade often proliferous from the face or margin.

Attached in soft mud, calcareous sand, or on rock, often among sea grasses.

Family Phyllosiphonaceae

Ostreobium quecketii Bornet *et* Flahault

Taylor (1960), p. 189; Humm and Taylor (1961), p. 373.

Slender, branched, coenocytic filaments, irregular in diameter and with occasional inflated portions; filaments 4-10 μm in diameter, tapering to 2 μm in diameter at the tips. Inflated portions to 40 μm in diameter.

Found in shells and calcareous tests of invertebrates; only observable after decalcification.

ORDER SIPHONOCLADALES

(to include the order Cladophorales *sensu* Dawes, 1969)

Family Cladophoraceae

Chaetomorpha aerea (Dillwyn) Kützing

Taylor (1960), p. 72; Humm and Taylor (1961), p. 352.

Filaments growing in dense clumps, each filament having a basal holdfast cell; cells 125-175 μm in diameter, sometimes to 300 μm or more. Base of filament usually more slender than the upper parts; the basal cell thick walled, about four times as long as wide at its greatest diameter, and producing rhizoidal outgrowths. Cells of the filament as long as broad to twice their diameter, the fertile cells swollen to nearly spherical.

Attached to rocks in shallow water; common at Point of Rocks.

Chaetomorpha brachygona Harvey

Taylor (1960), p. 70; Humm and Taylor (1961), p. 352.

Uniseriate, unbranched filaments; cells 80-175 μm in diameter, and about 1 diameter long.

Found in tide pools or shallow water of sheltered places, especially in salt marshes, where it becomes loose and entangled.

Chaetomorpha linum (Müller) Kützing

Taylor (1960), p. 71; Humm and Taylor (1961), p. 354.

Filaments similar to *C. brachygona,* except that they are larger (180-300 μm) in diameter; cells are usually a little longer than broad and thick walled.

Forming bright green mats in mangroves and salt marshes.

Chaetomorpha minima Collins *et* Hervey
Taylor (1960), p. 72.
Filaments attached by a discoid base and erect, to 5 mm tall, upper cells larger to 10-20 μm diameter and constricted at cross walls. Cells three diameters long, walls stratified.
Collected as epiphytes on *Thalassia* blades.

Cladophora catenata (Linnaeus) Kützing
Synonym: *Cladophora fuliginosa* Kützing
Taylor (1960), p. 83; Humm and Taylor (1961), p. 357; van den Hoek (1969*b*), p. 134.
Plants with main axes usually 150-160 μm in diameter; cells 5 to 10 diameters long. Filaments firm and stiff, dichotomous or alternately branched and bearing short, blunt branchlets in secund series.
Forming brownish green tufts, in shallow water. Parasitized by an endophytic fungus, *Blodgettia borneti* Wright.

Cladophora delicatula Montagne
Taylor (1960), p. 87; Humm and Taylor (1961), p. 354.
Plants soft, dull green, to 10 cm tall; main axis 40-60 μm in diameter, one to three diameters long.
Epiphytic in quiet water, especially marshes.

Cladophora fascicularis (Mertens) Kützing
Fig. 39
Taylor (1960), p. 91; Humm and Taylor (1961), p. 355.
Plants reaching 20 cm in height, branches fasciculate. Main axis 150-300 μm in diameter. Branchlets 70-100 μm in diameter, slightly tapered.
Common on rock walls and jetties, forming conspicuous green strands.

Cladophora frascatii Collins *et* Hervey
Taylor (1960), p. 82.
Plants forming mats, to about 2 cm tall; dichotomously branched although somewhat irregular; lower filaments around 100 μm in diameter.
Species conspicuous among other algae on rock jetties.

Cladophora glaucescens (Griffiths *ex* Harvey) Harvey
Taylor (1960), p. 86; Humm and Taylor (1961), p. 357.
Plants 7-30 cm or more in height, in loose tufts, much branched; the branches terminating in long, attenuate or sometimes secund ramelli. Main branches 40-60 μm in diameter; ramelli 20-30 μm, with cells usually three to four diameters long or longer.

Fig. 39. *Cladophora fascicularis.* A, *x 11*   B, *x 0.75*

Plants typically pale green, rather glassy and relatively rigid in spite of the slenderness of their filaments.

A common epiphyte.

Cladophora gracilis (Griffiths *ex* Harvey) Kützing

Taylor (1960), p. 90; Humm and Taylor (1961), p. 357.

Main axes less than 150 μm in diameter; axes irregularly bent and branching at the angles; upper parts bearing a secund series of long, tapering, acute branchlets, 40-60 μm in diameter and with cells three to five diameters long.

A most common species in salt marshes among *Spartina.* Littoral to sublittoral in distribution.

Cladophora longicellulata van den Hoek

Synonym: *Cladophora catenifera* Kützing

Taylor (1960), p. 92; van den Hoek (1969 *a*), p. 131.

Plants rigid; main axis 10 cm long (reaching 20 cm), distinct with opposite to trichotomous branching; angle of branching wide (45°-90°). Cells of main axis 150-170 μm wide, 350-420 μm long; cells of ultimate branches 170-240 μm wide, 230-320 μm long.

The plants were collected at Crystal River by dredge. They appear coarse and dark green, almost like a species of *Chara* or *Nitella* until examined with a microscope.

Cladophora repens (J. Agardh) Harvey
Taylor (1960), p. 82.
Forming soft mats on mud, filaments dichotomous, main axes 100-150 μm in diameter, cells two to three diameters long; cells in branches six to eight diameters long.
With *Boodleopsis* on mud, forming a continuous mat in red mangrove swamps; apparently aiding in stabilizing debris and substrate.

Cladophora scitula (Suhr) Kützing
Taylor (1960), p. 90.
Small, tufted plants, coarse; primary erect filaments 70-100 μm diameter, cells up to four diameters long; recurving branch and lateral branchlets pectinate and densely placed at tips; ultimate branches blunt to slightly conical.
Collected at Point of Rocks (Eiseman, in preparation).

Rhizoclonium hookeri Kützing
Taylor (1960), p. 77.
Plants with stiff filaments, entangled; branches to several cells in length; main axes of irregular diameter (40-80 μm), cells two to four diameters long.
Entangled with prop roots of red mangrove and in sea grass beds.

Rhizoclonium kochianum Kützing var. **kerneri** (Stockmayer) Hamel
Humm and Taylor (1961), p. 359.
Unbranched filaments, 10-14 μm in diameter; cells one to two diameters long, with a net-shaped plastid and pyrenoids.
Collected in Tampa Bay at the west end of Gandy Bridge (Humm and Taylor, 1961).

Rhizoclonium riparium (Roth) Harvey
Taylor (1960), p. 76; Humm and Taylor (1961), p. 359.
Similar to *R. kochianum* var. *kerneri* except that the filaments are usually 20-25 μm in diameter, one to two diameters long; it may produce short, rhizoidal branches (although rarely in Florida material).
A common species in salt marshes and brackish water.

Family Valoniaceae

Anadyomene menziesii Harvey
Taylor (1970), p. 125.
Plants with a massive rhizoidal base; one to many crisp, flat blades reaching up to 20 cm in length. Blades one cell thick and consist of two to six

parallel filaments that appear as veins. The area between the veins is filled with oval to elongate cells.

The species forms large leafy clusters and is common in depths of 30 m or more.

Anadyomene stellata (Wulfen) C. Agardh
Fig. 40
Taylor (1960), p. 125; Humm and Taylor (1961), p. 372.

Plants with a rhizoidal base and a crisp cluster of flat blades, 3-6 cm in height; the blades ovate to reniform in shape and one cell thick. Throughout the blade there are single, branching filaments which form pseudoveins. The area between the veins is filled with oval to elongate cells.

Present on pilings, limestone, and epiphytic on larger plants, such as *Digenia* and *Sargassum*.

A

B

Fig. 40. *Anadyomene stellata.* A, *x 2* B, *x 3.5*

Cladophoropsis membranacea (C. Agardh) Børgesen
  Taylor (1960), p. 117.
  A much branched uniseriate filament; forming a loose cushion or a coarse, dense tuft; without a main axis. Filaments 170-270 μm in diameter, branches 100-150 μm. Differing from *Cladophora* in the absence of cross walls at the base of branchlets.
  Epiphytic and on rock surfaces in the littoral and sublittoral zones.

Cystodictyon pavonium J. Agardh
  Taylor (1960), p. 124.
  Plants foliaceous, consisting of radiating filaments that appear as veins; cells 180-300 μm in diameter and 2 to 10 diameters long. Blade porous, the openings 0.5-10 mm in diameter. Plants somewhat similar to *Anadyomene* but having pores.
  A beautiful, rare, bright green porous blade; only found as fragments in depths of 40 m. Collected by dredge (Dawes and van Breedveld, 1969).

Struvea pulcherrima (J. A. Gray) Murray *et* Boodle
  Taylor (1960), p. 123.
  Plants to 25 cm in length, with two or three blades attached to the upper end of a stalk; stalk filament and filament of the blade of similar size (0.2-1 mm in diameter); blades form a network of filaments by anastomosing hapteral cells; attached to limestone by a complex rhizoidal system.
  Plants dredged off Tampa Bay in 20-80 m of water; they formed beautiful delicate blades up to 25 cm broad (Dawes and van Breedveld, 1969).

Valonia ventricosa J. Agardh
  Taylor (1960), p. 110.
  Plants consisting of one to a few cells. Cells oval to spherical in shape, up to 3 cm in diameter, and attached by rhizoidal projections from the lower end. Rhizoids are cut off from the main cell by cross walls.
  A number of small colonies collected off Sarasota in 30 m of water.

ORDER DASYCLADALES

Family Dasycladaceae

Acetabularia crenulata Lamouroux
  Fig. 41
  Taylor (1960), p. 105; Humm and Taylor (1961), p. 373.

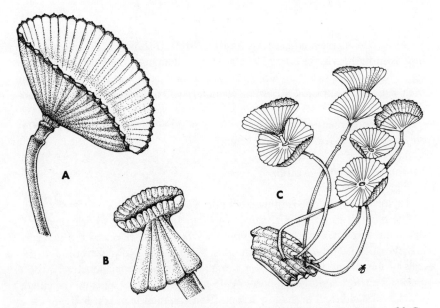

Fig. 41. *Acetabularia crenulata.* A, corona inferior, *x 4.5;* B, corona superior, *x 20;* C, habit sketch, *x 1.5.*

Plant composed of a slender, calcified stipe (7 cm tall) bearing a concave disk at the top. The disk is 6-18 mm in diameter and composed of 30-80 moderately calcified segments (rays). Some plants produce two or more disks, one above the other. A crown of projections (corona superior), arises from the center, each segment of which bears a colorless hair. A similar structure (corona inferior) is located on the underside of the disk, but it lacks hairs. Many spores (200 to 500) are borne within the rays.

Plants in dense clusters on shells and stones in shallow water, sometimes among sea grasses.

Acetabularia farlowii Solms-Laubach

Taylor (1960), p. 105; Humm and Taylor (1961), p. 374.

Differs from *A. crenulata* in having a smaller disk (4-7 mm in diameter), shorter stipes (1-2 cm high), and fewer rays (20 to 30) which are usually separate. Two hair scars on each segment of corona superior; spores 40 to 120 per ray.

Grows attached to shells and stones in shallow water, often in association with the more abundant *A. crenulata.*

Acicularia schenckii (Möbius) Solms-Laubach

Taylor (1960), p. 107; Humm and Taylor (1961), p. 375.

Similar to *A. crenulata*, except that the stipe is shorter (1-3 cm high), the disk is more heavily calcified, smaller (6 mm diameter), and there are fewer rays (30 to 50). Corona superior, 13 mm in diameter, with two hair scars on each segment. Spores 60-80 μm in diameter, 100 to 200 per ray, and embedded in lime.

Uncommon plant on rocks or shells in offshore areas (2-3 m). It is considered a species of *Acetabularia* by some.

Batophora oerstedi J. Agardh
Fig. 42

Taylor (1960), p. 98; Humm and Taylor (1961), p. 375.

Plants with an erect axis, usually 4-6 cm high; bearing whorls of monosiphonous branches that are dichotomously branched six or seven times; the branches 0.5-1.0 mm apart on the axis. Green, oval sporangia borne at the ends of the second to fourth series of branches; the sporangia mostly 700-850 μm in diameter; containing spores 40-45 μm in diameter. Plants not calcified.

On rocks and shell in shallow water, often common in brackish areas.

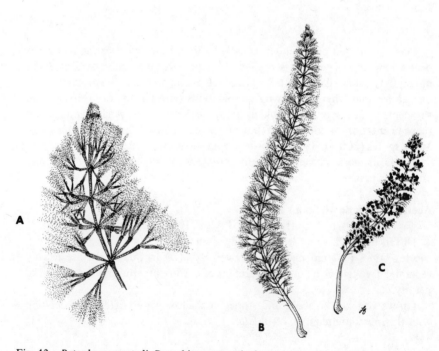

Fig. 42. *Batophora oerstedi.* Branching pattern is shown in A, *x 10;* habit sketch is seen in B, *x 1.75;* and plant with gametangia is seen in C, *x 1.25.*

Class Xanthophyceae

ORDER HETEROSIPHONALES

Family Vaucheriaceae

Vaucheria thuretii Woronin
Humm and Taylor (1961), p. 376.

Plant of branched coenocytic filaments producing a small green cushion. Filaments to 80 μm in diameter; antheridia sessile, ovoid in shape, 50-70 μm in diameter; oogonia spherical to pyriform, on a one-celled stalk, about 200 μm in diameter.

The plant forms matted growth on mud or other surfaces in salt marshes and mangrove swamps; in littoral mud flats (Dawes, 1965).

6

Phaeophyta
SPECIES DESCRIPTIONS

Taxonomic literature and other figures for the species are available in Earle (1969) and Taylor (1960).

ORDER ECTOCARPALES

Family Ectocarpaceae

Acinetospora crinita (Carmichael) Kornmann

Taylor (1960), as *A. pusilla* (Griffiths) Bornet p. 214; Earle (1960), p. 128.

Plants filamentous, in tufts 4-8 cm tall, often entangled below, with the tips usually free and flexuous. Cells 15-56 μm in diameter, one to six diameters long, with discoid plastids. Growth intercalary; branching irregular or opposite, three to five cells long. Branches arising at right angles to the main axis, lateral rhizoids common. Monosporangia (unilocular sporangia?) sessile or short stalked, subspherical, 31-42 μm by 32-40 μm.

In the littoral zone on rocks and sea walls where there is wave action. Occurs during the spring and summer months. In late summer it may be found entangled with other algae and drifting.

Bachelotia antillarum (Grunow) Gerloff

Taylor (1960), as *Pylaiella antillarum* (Grunow) De Toni (p. 197) and *Bachelotia fulvescens* (Bornet) Kuckuck (p. 198); Earle (1969), p. 129.

Plants filamentous, 1.5-3.5 cm tall, with creeping and erect portions; the stoloniferous part giving rise to erect filaments and anchoring haptera. Cells 11-47 μm in diameter, shorter than wide or up to three diameters long. Plastids stellate, one or two in each mature cell; at times the plastids appearing discoid in other parts of the plant, especially in embryonic and degenerate cells. Unilocular sporangia in series, usually intercalary in the branches.

Generally occurs on rocks and shells and is a spring inhabitant of Tampa Bay.

Ectocarpus dasycarpus Kuckuck

Taylor (1960), p. 200; Earle (1969), p. 132.

Plants filamentous, tufted, 5-7 cm tall; pseudodichotomously branched and with small lateral branchlets. Mature cells 20-40 μm in diameter and two to three diameters long. With forked or band-shaped plastids. Plurilocular sporangia usually terminal on one to several-celled lateral branches; sporangia are 10-15 μm in diameter and up to 250 μm in length.

Plants from Tampa Bay have axes mostly 28 μm in diameter with sporangia 18 μm in diameter and 180 μm or more in length. Grows as an epiphyte in areas where salinities are below 25 0/00. (Salinities are measured in grams of salt per kilogram of water and given in parts per thousand.) Found only in the winter months in this region.

Ectocarpus elachistaeformis Heydrich

Taylor (1960), p. 202; Earle (1969), p. 133.

Plant a minute, filamentous epiphyte; basal portion with decumbent or penetrating simple to branched haptera. Erect filaments 1-1.5 mm high, simple or basally branched, sometimes hair tipped; mature cells 10-18 μm in diameter, one to three diameters long. Plurilocular sporangia fusiform or with a tapering apex, measuring 15-25 μm by 63-80 μm or more, and usually located near the base of erect filaments.

Epiphytic or entangled on coarse algae and sea grasses, as well as such artificial substrate as buoys and plastic fishing lines. Present in Tampa Bay in autumn and winter.

Ectocarpus intermedius Kützing

Taylor (1960) as *E. confervoides* (Roth) Le Jolis, p. 200; Earle (1969), p. 130.

Plants filamentous, 5-50 cm high, attached by basal rhizoids or becoming free-floating. Branches usually alternate, tapering, sometimes hair tipped. Mature cells 20-50 μm by 35-60 μm, with ribbonlike plastids. Sporangia oval, sessile or short stalked, fusiform; 10-40 μm by 60-150 μm (occasionally to 400 μm).

Grows on various substrate, including rocks, buoys, wood, sea grasses, and algae. Present from littoral zone to moderate depths in a wide range of salinities.

Ectocarpus siliculosus (Dillwyn) Lyngbye
Taylor (1960), p. 199; Earle (1969), p. 133.

Plant filamentous, up to three dm tall, attached or free-floating; branching alternate or unilateral. Cells of the main axes 40-60 μm in diameter, four or five diameters long when mature, though shorter in younger portions. Unilocular sporangia short stalked or sessile, ellipsoid, and measuring 20-27 μm by 30-65 μm; often on the same plant with plurilocular sporangia. Plurilocular sporangia are also sessile or short stalked, typically conical, 12-25 μm by 50-600 μm, and usually terminating in a hair.

Typically found in winter, growing on various substrate and under a broad range of salinity conditions.

Giffordia conifera (Børgesen) Taylor
Taylor (1960), p. 207; Earle (1969), p. 135.

Plants filamentous, with erect axes 5-8 cm tall, attached by basal rhizoids; branches alternate, usually hair tipped. Cells 40 μm in diameter, one-half to four diameters long, with discoid plastids. Unilocular sporangia oval, usually solitary in the axils of the branchlets. Plurilocular sporangia conical, usually occurring singly or in a series along the branches, 24-40 μm by 40-110 μm, and close to the main axis.

In shallow water, usually epiphytic on *Thalassia*. Primarily found in the spring and summer. It has been dredged from a depth of 25 m. Tampa Bay is the northernmost record for the species in the Gulf of Mexico.

Giffordia indica (Sonder) Papenfuss *et* Chihara
Taylor (1960), as *G. duchassaigniana* (Grunow) Taylor, p. 207; Earle (1960), p. 136.

Plants soft, filamentous with a rhizoidal base and irregularly branched; hair-tipped filaments 2-5 cm tall. Cells 20-34 μm in diameter, one-half to one and one-half diameters long and with discoid plastids. Unilocular sporangia sessile, oval, 70 μm in diameter and up to 110 μm long; plurilocular sporangia sessile or short stalked, usually cylindrical and with a blunt apex, 20-50 μm by 100-250 μm in size.

Common in the summer on limestone, oyster bars, sponges, and sea grasses; to depths of 20 m.

Giffordia mitchelliae (Harvey) Hamel
Taylor (1960), as *G. mitchellae*, p. 206; Earle (1969), p. 138.

Plants filamentous, in masses 2-3 cm tall to several decimeters in length; attached by basal rhizoids or detached. Branching abundant, usually alternate; branches sometimes hair tipped; branchlet cells 15-20 μm in diameter; cells of the main axis 35-50 μm in diameter, two to three diameters long; all cells with discoid plastids. Unilocular sporangia sessile, oblong, 25-45 μm by 60-80 μm; plurilocular sporangia elongate to cylindrical, 20-30 μm by 50-150 μm, and with a blunt apex. Two types of plurilocular sporangia are known; meiosporangia with cells 6-7 μm in diameter, and megasporangia with cells 10-17 μm in diameter.

A conspicuous species in the summer and autumn in Tampa Bay, forming loose masses of filaments entangled among other algae and sea grasses.

Giffordia rallsiae (Vickers) Taylor

Taylor (1960), p. 208; Earle (1969), p. 140.

Plants filamentous, with an erect portion to 5 mm and a creeping base attached by rhizoids. Branches occasional, sometimes hair tipped, with cells 27-40 μm in diameter and two to five diameters long. Unilocular sporangia oval, 40-45 μm by 70 μm; plurilocular sporangia sessile or stalked, fusiform, and with a tapered to acute apex, 27-40 μm by 80-120 μm; located near the base of a branch.

Epiphytic on *Codium* and *Sargassum,* infrequent in Tampa Bay in the fall and winter months.

ORDER SPHACELARIALES

Family Sphacelariaceae

Sphacelaria furcigera Kützing

Taylor (1960), p. 210; Earle (1969), p. 144.

Plants forming tufts with a creeping or penetrating basal portion and erect filaments; 1.5-3.0 cm tall. Cells 16-45 μm in diameter, as long as broad or slightly longer, occasionally with longitudinal divisions (pluriseriate). Unilocular sporangia spherical, 50-70 μm in diameter; plurilocular sporangia of two types, one 45-65 μm by 24-48 μm, with locules about 3 μm in diameter; the other 30-60 μm by 28-40 μm, with locules about 6 μm in diameter. Propagulae biradiate originating from a slender base, with base and arms approximately equal in length; propagulae 24 μm in diameter with the distance between the tips up to 450 μm.

Plants epiphytic and on buoys, shells, and rocks in water up to 20 m deep, occurring year-round in Tampa Bay.

Sphacelaria tribuloides Meneghini

Taylor (1960), p. 211; Earle (1969), p. 146.

Plants in tufts 4-5 mm tall, with pluriseriate filaments 25-60 μm in diameter. Cells as long as broad or slightly longer. Propagulae biradiate from a broad base, shield shaped or broadly triangular in side view, with a span between the arms of 140-165 μm.

Common on limestone and other hard surfaces as well as on coarse algae and sea grasses.

ORDER DICTYOTALES

Family Dictyotaceae

Dictyopteris delicatula Lamouroux
    Taylor (1960), p. 227; Earle (1969), p. 149.

    Plants erect, 2-8 cm tall, with flat, dichotomous to irregular branches; the branches 0.5-5.0 mm broad, with a pronounced midrib, and in mature blades, inconspicuous marginal ribs; apices rounded. Blades distromatic, with cells in distinct divaricate rows, but not forming pinnate veinlets. Sori in rows along the midrib.

    Collected offshore in the Tampa Bay area to depths of 20 m, on limestone, but it occurs in shallow water in the Florida Keys. Fragments tentatively identified as another species (*D. membranacea*) have been found in offshore dredges only once.

Dictyota cervicornis Kützing
    Taylor (1960), p. 222; Earle (1969), p. 153.

    Plants erect, bushy, to 2 dm tall, with flat, subdichotomous to asymmetrical branches; 1.0-2.5 mm broad, with acute tips.

    Collected by dredging (10-20 m) offshore in Tampa Bay.

Dictyota ciliolata Kützing
    Taylor (1960), p. 223; Earle (1969), p. 155.

    Plants erect, branching dichotomous; branch width 7-12 mm with widest region at forkings; margins aculeate-dentate and projections regularly spaced.

    Specimens were found in May off Anclote Key in 10 m of water. Earle discusses this species and its taxonomic relationship to *D. dichotoma*.

Dictyota dentata Lamouroux
    Taylor (1960), p. 224; Earle (1969), p. 156.

    Plants erect, bushy, 1-2 dm tall, with well-defined main axes; branches alternate, 2-3 mm in diameter; with some spurlike branchlets 1-2 mm long. Tips round or acute.

    Found in drift in autumn; known south of Tampa in depths up to 15 m.

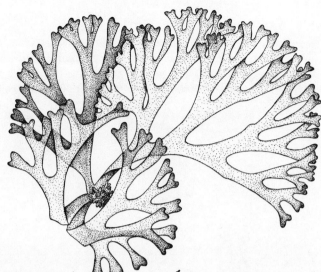

Fig. 43. *Dictyota dichotoma.* x 1

Dictyota dichotoma (Hudson) Lamouroux
Fig. 43
Taylor (1960), p. 218; Earle (1969), p. 157.

Plants erect, to 3.5 dm tall, although usually smaller; with strap-shaped, sometimes twisted, dichotomous branches 2-5 mm in diameter. Branches form angles of 15° to 45°; tips blunt to somewhat tapered.

Occurring in shallow water and to depths of 5-8 m, growing on a variety of substrate. The species is the most widespread of this genus with morphological forms, some of which are considered to be separate species (see S. E. Taylor, 1966). *Dictyota indica* Sonder is one such example in which the major difference is the length of internodes (1.5-4 cm long) and the width of the frond (1.0-1.5 mm broad). Material identifiable as *D. indica* was collected offshore of Anclote Key.

Dictyota divaricata Lamouroux
Fig. 44
Taylor (1960), p. 221; Earle (1969), p. 160.

Plants erect, often entangled, 3-7 cm high; with flat, dichotomous branches about 3 mm broad in the lower portion; fronds narrowing abruptly to 0.1-0.2 mm in the upper portions; branches forming angles of 90° to 120°; tips usually blunt or rounded.

Common year-round, attached to limestone or sea grasses.

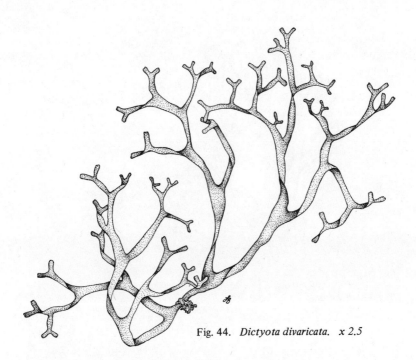

Fig. 44. *Dictyota divaricata.* x 2.5

Dictyota linearis (C. Agardh) Greville
　　Taylor (1960), p. 219; Earle (1969), p. 161.
　　Plants erect, 5-12 cm tall, branches 0.5-1.0 mm broad; branching dichotomous to slightly irregular. Distance between dichotomies (internodes) 0.5-2.0 cm; angle of branching narrow; tips acute.
　　'The plant is distinguished by its slender dichotomous blades, but there are forms between this and *D. dichotoma, D. divaricata,* and *D. cervicornis.* Thus, the validity of *D. linearis* is doubtful (Earle, 1969).
　　A plant of shallow water growing on rocks, seawalls, or other algae; apparently a spring and summer plant in the Tampa Bay region.

Dilophus alternans J. Agardh
　　Taylor (1960), p. 216; Earle (1969), p. 163.
　　Plants 1.0-1.5 dm tall, bushy, with alternate branching, the final branching dichotomous. Branches strap-shaped 0.5-3.0 mm in diameter, the axes 2-4 mm in diameter; tips obtuse. Mature blades with a medulla of two or more cells at the margins, covered by a one-celled cortical layer.
　　Reported from reefs offshore Tampa Bay; the known northern range of the species is the Gulf of Mexico (Phillips and Springer, 1960).

Lobophora variegata (Lamouroux) Womersley
　　As *Pocockiella variegata* (Lamouroux) Papenfuss, Taylor (1960), p. 231; Earle (1969), p. 173.

Plants decumbent or erect, with flat, usually deltoid or suborbicular blades; blades 1-7 cm wide, often in dense, overlapping clusters; attached by rhizoids. In surface view, cells of blade in vertical rows; in cross section, the blades 100-300 μm thick, with a one-layered medulla and a subcortex of two or more layers of cells. Fertile sori scattered over both surfaces.

Common in the Florida Keys and found occasionally on limestone rocks offshore Homosassa River.

Padina vickersiae Hoyt
Fig. 45
Taylor (1960), p. 236; Earle (1969), p. 172.

Plants with flat, fan-shaped blades, having an involute margin; blades forming clusters 10-15 cm tall, blades entire, split, or proliferous. Plants having little or no calcification and attached by a rhizoidal holdfast. Blades two cells thick (50 μm) near the margins, but four to six or sometimes eight cells (150-220 μm) in the lower portions. Bands of hairs zonate, hairs 2-7 mm in diameter, the zones usually about 4 mm apart and alternating with fertile regions. Sporangia 100-200 μm in diameter and covered by an indusium. Plants

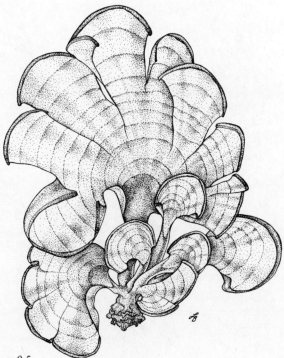

Fig. 45. *Padina vickersiae.* x 0.5

dioecious, with antheridia 10-20 μm in diameter and oogonia 30-65 μm in diameter.

Occurring from lower littoral areas to depths of at least 14 m and found attached to hard substrate. Male and female plants found in the fall (November to December), while sporophytes occur year-round (Fagerberg, 1972).

Spatoglossum schroederi (C. Agardh) Kützing
   Fig. 46
   Taylor (1960), p. 225; Earle (1969), p. 176.
   Plants erect, 1-3 dm tall, with alternate, subdichotomous to palmate branches or lobes; divisions 0.5-2.5 cm in diameter, with undulate or dentate

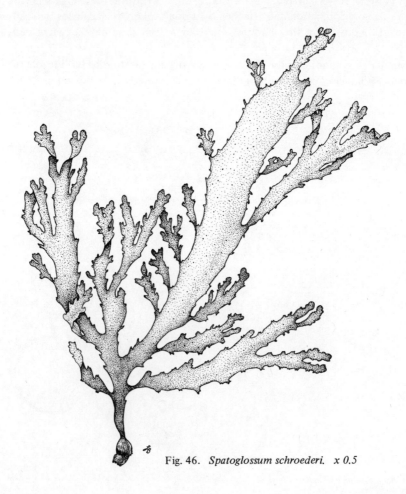

Fig. 46. *Spatoglossum schroederi. x 0.5*

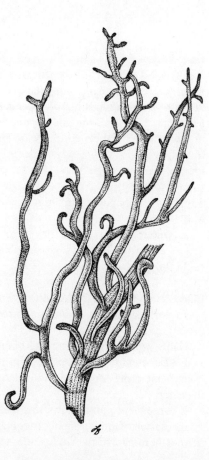

Fig. 47. *Cladosiphon occidentalis.* A number of plants are shown as epiphytes on the blade of *Thalassia.* x 2.5

margins; in cross section the thallus having large central cells surrounded by a thin monostromatic cortical layer. Growth is from a series of apical cells.

In the Tampa Bay area, this species has been found only on offshore reefs in 10-20 m of water.

ORDER CHORDARIALES

Family Chordariaceae

Cladosiphon occidentalis Kylin
Fig. 47
Taylor (1960), p. 248; Earle (1969), p. 180.
Plants light to dark brown in color, of soft, gelatinous slippery texture, cylindrical and filiform, nearly simple to much branched, 1-3 dm tall, at-

tached by a small basal disk. Axis of filaments organized into a hollow or basally solid cylinder, filaments embedded in mucilage; branches 0.75-3.0 mm wide, formed irregularly along the axis. Medullary filaments, 50-112 μm wide, 200-400 μm long, bearing peripheral filaments 100-300 μm high, and colorless hairs, 8-15 μm wide and up to 150 μm long. Unilocular sporangia single or in groups of three at the base of the peripheral pigmented filaments, oblong-ovate in shape, when mature 30-40 μm wide and 50-80 μm long. Plurilocular sporangia may also be present with unilocular sporangia, the former at the tips of peripheral pigmented filaments; plurilocular sporangia 13-19 μm in diameter, uniseriate or pluriseriate and branched, up to 48 μm long.

A winter species, occurring throughout Tampa Bay; epiphytic on sea grasses and other hosts. Able to withstand a wide range of salinities (23 0/00 to 36 0/00).

Cladosiphon zosterae (J. Agardh) Kylin

Taylor (1960), as *Eudesme zosterae* (J. Agardh) Kylin, p. 247; Earle (1969), p. 182.

Plants gelatinous, filiform and branched; usually epiphytic on sea grasses. Like *Cladosiphon occidentalis* except that the medullary filaments are loosely bound and easily separated. Assimilatory filaments wider, up to 20-40 μm in diameter.

If subsequent study shows this species to be distinct from *C. occidentalis* in the eastern Gulf of Mexico, probably its distribution will be primarily in the northern part of the Gulf (Earle, 1969).

Family Myrionemataceae

Ascocyclus orbicularis (J. Agardh) Magnus

Taylor (1960), p. 242; Earle (1969), p. 178.

Plants epiphytic, forming light brown patches, 1-3 mm or more in diameter; composed of radiating prostrate filaments and upright stalked paraphyses. Paraphyses ovate when first formed, later elongate, 8-25 μm in diameter and up to 170 μm long. Plurilocular sporangia sessile or on one-celled stalks, cylindrical, obtuse, uniseriate, 5-12 μm by 20-75 μm.

Epiphytic on sea grasses and other algae to 2 m depth. A winter and spring species in the Tampa Bay vicinity.

Myrionema strangulans Greville

Taylor (1960), p. 240; Earle (1969), p. 179.

Plants forming minute disks composed of loose or closely associated filaments; filaments 50-100 μm long, cells 6-11 μm in diameter and one to three diameters long; having discoid plastids in mature cells; hairs 8-13 μm in diameter, present among the filaments. Unilocular sporangia located basally, ellipsoid or obovoid in shape, 20-35 μm by 35-65 μm; plurilocular sporangia sessile or on short stalks, bluntly cylindrical, 7-11 μm by 15-50 μm.

A small epiphyte of larger algae and sea grasses in shallow water.

Family Spermatochnaceae

Nemacystus howei (Taylor) Kylin
   Taylor (1960), p. 249; Earle (1969), p. 184.
   Plants dark brown, of soft and gelatinous texture, cylindrical and filiform, 2-4 dm tall with numerous short, alternate or irregular branches, 0.3-1.0 mm in diameter. Axis a solid cylinder of filaments, apparent in young portions; cells of the medulla 58-166 μm by 200-1,400 μm. Assimilatory filaments unbranched, borne directly upon medullary cells; cells of the filaments 10-16.5 μm in diameter, with about eight rounded cells to each filament. Unilocular sporangia located at the base of lateral filaments, spherical to oval in shape, 18-37 μm by 20-45 μm, usually several together; plurilocular sporangia in the same position, occurring in groups up to 10 in a cluster (usually 1 to 2); 6.5-8.5 μm by 36.5-78.0 μm.

Collected offshore from Tampa Bay in 13 m of water and offshore Homosassa River in 3 m of water. Common in the spring.

Family Stilophoraceae

Stilophora rhizodes (Turner) J. Agardh
   Fig. 48
   Taylor (1960), p. 250; Earle (1969), p. 185.
   Plants with slender, cylindrical, dichotomous to irregularly branched axes, 1-3 dm tall, often tapering to very slender or threadlike recurved tips; attachment by basal disk. Medulla of young plants solid, older portions often hollow; having intercalary (trichothallic) growth. Sori with colorless hairs and curved filaments; the filaments are 3.5 μm in diameter at the base to 9.5-13.0 μm in diameter near the tips, 75-85 μm long; sori appearing as dark, round protuberances on the branches. Unilocular sporangia scattered, located at the bases of the clustered filaments, club-shaped, 22-32 μm by 36-56 μm; plurilocular sporangia on separate plants, uniseriate and composed of four to nine cells, 9.5-11.5 μm by 30-50 μm.

A winter-spring species in the Gulf, reaching its southern known limit at Tampa Bay. Common epiphyte on sea grasses and algae.

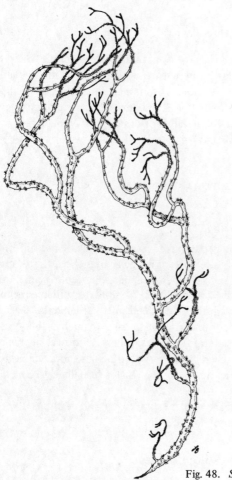

Fig. 48. *Stilophora rhizodes.* x 2

ORDER SPOROCHNALES
Family Sporochnaceae

Sporochnus bolleanus Montagne
   Taylor (1960), p. 253; Earle (1969), p. 190.
   Plants with dark brown, wiry, filiform axes, up to 1 m tall; branching alternate to irregular; branchlets terminating in tufts of light olive brown filaments; the branchlets with fertile portions up to 3.6 mm long; the branchlets on pedicels 1.5-10 mm long; filaments of the young branches in tufts 10-15 mm in length.
   Occurring during the spring in shallow water; growing on a variety of substrate, offshore of Homosassa Springs.

Sporochnus pedunculatus (Hudson) C. Agardh
Fig. 49
Taylor (1960), p. 253; Earle (1969), p. 190.

Plant with dark brown, wiry, filiform axes, to 3 dm tall or more; with many regular branches up to 2 dm long, bearing numerous tufts of light olive brown filaments on branchlets. Branchlets bearing tufts of hairs with fertile portions 0.75-1.25 mm (rarely 2.0 mm) long, 0.5 mm in diameter, pedicels of branchlets 1-2 mm long; when plants are young the filaments in tufts 3-5 mm long, dehiscent in older plants. Unilocular sporangia uniformly distributed on

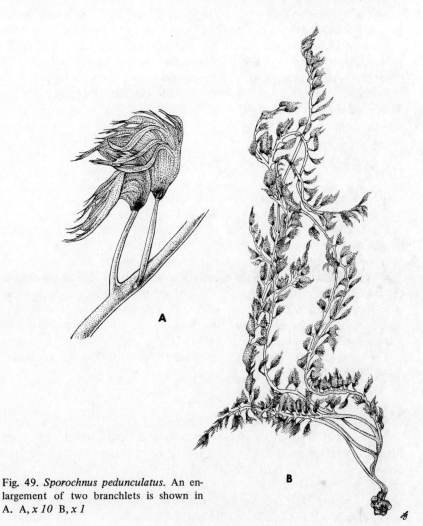

Fig. 49. *Sporochnus pedunculatus.* An enlargement of two branchlets is shown in A. A, *x 10* B, *x 1*

short filaments at the base of the long, conspicuous filaments.
A winter and spring species along the west coast of Florida.

ORDER DICTYOSIPHONALES
Family Stictyosiphonaceae

Stictyosiphon subsimplex Holden
  Earle (1969), p. 193.
  Plants dark brown, filiform, usually 3-8 cm high, 0.5-1.0 mm wide, simple
or with occasional branches; attached by a rhizoidal holdfast. Axis solid,
round in cross section, with four large cells surrounded by a cortex of smaller
cells; unilocular sporangia and colorless hairs formed on the surface of the
plant; unilocular sporangia 40-56 μm in diameter, and 56-84 μm long.
  Common on *Ruppia* and *Diplanthera* in littoral, estuarine areas during the
winter and spring months.

Family Punctariaceae

Colpomenia sinuosa (Roth) Derbés *et* Solier
  Taylor (1960), p. 260; Earle (1969), p. 200.
  Plants golden brown, spherical to lobed, hollow, often 6-10 cm across, but
attaining 20 cm; composed of a layer of cells to 0.5 mm thick, the inner cells
colorless, to 180 μm in diameter and grading in size to small surface cells
3.7-7.5 μm in diameter. Groups of colorless hairs surrounded by plurilocular
sporangia are found scattered over the surface; the plurilocular sporangia are
cylindrical to clavate, 3.7-7.5 μm wide and 18-30 μm long, occasionally associated with club-shaped paraphyses. Sori appear as dark, raised patches
1.0-1.5 mm in diameter.
  Growing on limestone and as an epiphyte; dredged from 15 m offshore
Tampa Bay in the spring.

Myriotrichia occidentalis Børgesen
  Taylor (1960), p. 257; Earle (1969), p. 195.
  Filamentous epiphytes about 1 mm tall; with basal filaments irregularly
branched and creeping, not forming a disk; erect filaments becoming pluriseriate, cells 18 μm in diameter and to 18 μm long; erect filaments often hair
tipped, the hairs about 12 μm in diameter. Plurilocular sporangia sessile to
pedicellate. Unilocular sporangia spherical, 18-20 μm diameter.
  Specimens reported from northwestern Florida (Earle, 1969) and found
off Homosassa River.

Myriotrichia subcorymbosa (Holden) Blomquist

Earle (1969), p. 197.

Plant filamentous, 1-3 mm tall, with a discoid base of laterally joined filaments; erect portions with alternate or opposite branches; cells 12-15 μm in diameter, 2-4 diameters long and with disk-shaped plastids. Hairs terminal or lateral, sometimes opposite a branch. Plurilocular sporangia terminal or lateral, often opposite or clustered, sessile or upon a stalk of one to three cells; plurilocular sporangia cylindrical, 18-25 μm in diameter, 30-140 μm long, with few longitudinal divisions (rarely pluriseriate).

Myriotrichia is epiphytic on other algae and on sea grasses, found in depths to 8.5 m, apparently the entire year.

Rosenvingea intricata (J. Agardh) Børgesen

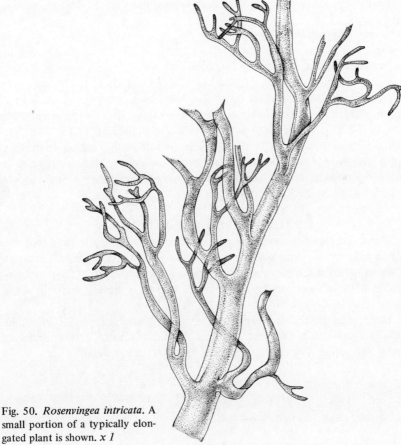

Fig. 50. *Rosenvingea intricata*. A small portion of a typically elongated plant is shown. *x 1*

Fig. 50

Taylor (1960), p. 262; Earle (1969), p. 207.

Plants bushy, usually much branched, golden to olive brown in color, to 4 dm tall, hollow and attached by a basal disk. Branches flattened, sometimes contorted, sharply narrowing from the main axes to the ultimate branchlets, with tips usually filiform and attenuate. Cells in surface view irregular or with a linear arrangement. Plurilocular sporangia in elliptical sori.

Collected by dredging offshore from Tampa Bay. *Rosenvingea orientalis* (J. Agardh) Børgesen has been recorded by Dawes *et al.* (1967) and Earle (1969) south of Cape Romano, Florida, and it has been collected in drift by Eiseman (in preparation) at Point of Rocks.

ORDER FUCALES

Family Sargassaceae

Sargassum filipendula C. Agardh

Fig. 51

Taylor (1960), p. 270; Earle (1969), p. 217.

Plants attached by a large lobed holdfast, with smooth, terete, and erect and alternately branched axes up to 2 m high, axes commonly 0.5-1.0 m in length. Blades lanceolate foliar with slender bases. Blades thin, serrate, 5-8 mm by 3-8 cm, simple or sometimes forked near the base of the plant and with a distinct midrib; cryptostomata scattered or in one or more rows along the midrib. Air bladders spherical, sometimes apiculate, bladders 3-5 mm in diameter, on stalks about 5 mm long; receptacle on an axillary branch, simple or occasionally branched.

Var. **montagnei** (Bailey) Collins *et* Hervey

Taylor (1960), p. 271; Earle (1969), p. 220.

Blades long, slender to linear, 1-5 mm wide, 2-7.5 (up to 15) cm long, tapering to apex, with a distinct midrib; cryptostomata present; margins nearly smooth to serrate. Vesicles (bladders) round to oval, 2-3 mm in diameter, sometimes apiculate; on stalks one-half to three times the length of the vesicle; receptacles cylindrical.

The typical form and variety are commonly found on natural limestone. Although present throughout the year, growth is most luxuriant during winter and spring. The variety is usually found in deeper water.

Sargassum fluitans Børgesen

Taylor (1960), p. 281; Earle (1969), p. 221.

Plants pelagic, usually golden brown in color, with smooth or slightly

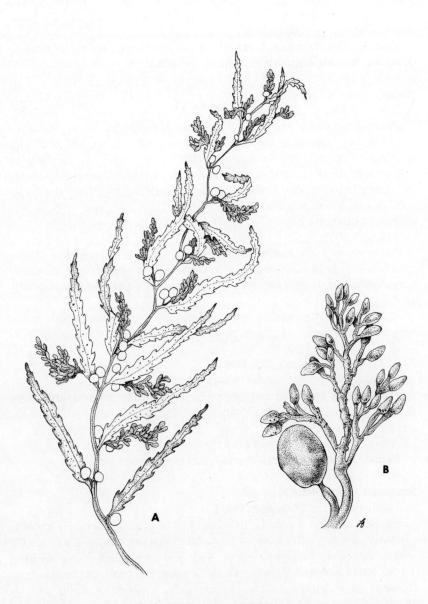

Fig. 51. *Sargassum filipendula*. A, a habit sketch of the upper portion of the plant, *x 0.5*; B, an enlargement of a fertile branch (with receptacles), *x 4*

spiny axes bearing firm, short-stalked, lanceolate, broadly serrate blades. Blades 2-6 cm long, 3-8 mm in diameter, usually lacking cryptostomata. Vesicles round to oval, not apiculate, 4-5 mm in diameter on stalks 2-3 mm in length. Only sterile plants known.

Drifting ashore sporadically, usually in association with the more abundant *S. natans.* The plant appears sensitive to low salinities.

Sargassum hystrix J. Agardh

Taylor (1960), p. 279; Earle (1969), p. 223.

Plants attached by a basal holdfast; with axes smooth, slender, to 44 dm tall. The axes bearing closely set, radially disposed blades; blades oblong-elliptical in shape, with serrate margins, asymmetrical bases, conspicuous midrib, and small, scattered cryptostomata. Air bladders round, usually numerous, borne on short stalks; receptacles terete, on branchlets about one-third as long as blades (palmately divided).

Var. **buxifolium** Chauvin

Earle (1969), p. 225.

Plants up to 5 dm high with numerous short, lateral branchlets. Blades crowded, 1.5 cm in diameter, to 6 cm long, oval, or oblong, with smooth or very slightly serrate margins; cryptostomata few, inconspicuous. Plant usually sterile, often found drifting.

Occurrence of this species in the Tampa Bay area is questionable; it has been reported from offshore reefs by Phillips and Springer (1960).

Sargassum natans (Linnaeus) Gaillon

Taylor (1960), p. 281; Earle (1969), p. 225.

Plants widely branched and tangled, pelagic in habit. Stipe smooth, with firm linear blades, 2.5-7.0 (or to 10) cm long and 2.0-3.5 mm in width, blades serrate. Cryptostomata absent, receptacles unknown; vesicles apiculate and sometimes with a very long spine 3-5 mm in diameter.

Similar to *S. fluitans* in distribution, but more abundant.

Sargassum polyceratium Montagne

Taylor (1960), p. 276; Earle (1969), p. 227.

Plants erect with a disklike holdfast, axes muriculate to nearly smooth; plant 4-9 dm tall, ultimately with spurlike branches; blades nearly sessile, serrate, lanceolate to broadly ovate with a flat or broadly rounded base; 1.5-3.5 cm in length, 5-10 mm wide, with numerous small, scattered crypto-stomata. Air bladders spherical, 3-6 mm wide, sessile or on stalks 2-6 mm long; receptacles on short, forking axillary branches.

Found in drift in Tampa Bay. Apparently restricted to warm waters and high salinities offshore.

Sargassum pteropleuron Grunow
Fig. 52
Taylor (1960), p. 274; Earle (1969), p. 228.

Plants pale gold to medium brown in color, attached by a large holdfast, with terete to angular muriculate axes, 3-10 dm tall (sometimes more). Side branches bearing narrow to linear blades having a pronounced midrib, which

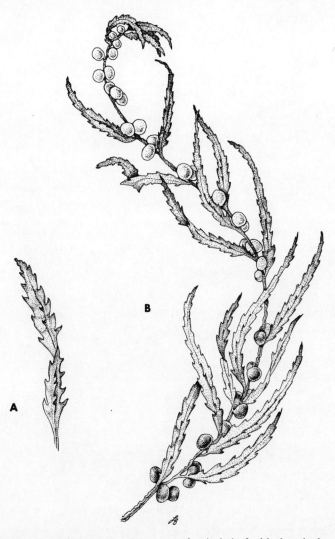

Fig. 52. *Sargassum pteropleuron.* An enlargement of a single leaf with the raised costa (midrib) is shown in A.   A, *x 0.5*   B, *x 0.5*

often develops a prominent costate (raised) ridge or wing; blades broad, 2.5-7.5 mm in diameter, sometimes broader near the base. Air bladders rounded, 3-8 mm in diameter, nearly sessile or on stalks 2-3 mm long. Blades may become prismatic in cross section.

A common shallow water (1-5 m) plant occurring on rocky substrate under a wide range of salinities (10 0/00 to 30 0/00) and temperatures (15°-36°C); found throughout the year.

7

Rhodophyta

SPECIES DESCRIPTIONS

Taxonomic literature and other figures can be found in Taylor (1960).

ORDER BANGIALES

Family Bangiaceae

Asterocystis ramosa (Thwaites) Gobi
Taylor (1960), p. 287.
Filaments in soft tufts, 12-30 μm in diameter, with irregular branching, uniseriate, cells 9 μm in diameter and up to 20 μm long. Akinetes uncommon but present, up to 16 μm diameter.
Epiphytic on sea grass blades in vicinity of Anclote River (Ballantine, 1972).

Erythrocladia subintegra Rosenvinge
Taylor (1960), p. 290.
Epiphytic, forming disks to about 50 μm in diameter. Filaments branching by division of apical cells (cells on edge of disk). Cells 4-6 μm in diameter.
Epiphytic on sea grasses and coarse algae, common from May to June.

Erythrotrichia carnea (Dillwyn) J. Agardh
Taylor (1960), p. 292.
Plants consisting of uniseriate filaments or with a few longitudinal walls.

Filaments attached by a basal cell that is lobed or produces short rhizoids. Cells somewhat spindle shaped to swollen and 16-30 μm in diameter.

An occasional epiphyte on sea grasses and various other algae.

Goniotrichum alsidii (Zanardini) Howe

Taylor (1960), p. 288.

Thallus of microscopic erect tufts of freely branching filaments (0.3-0.6 mm in diameter). Cells 7-13 μm in diameter and 4-13 μm long. Plastids violet to rosy red.

Abundant along the west coast of Florida, on sea grasses and coarse algae.

ORDER NEMALIONALES

Family Acrochaetiaceae

Acrochaetium avrainvilleae Børgesen

Taylor (1960), p. 311.

Plants to about 1 mm, little branched, cells of main axes about 9 μm in diameter, 33 μm long. Cells with one large "pyrenoid." Basal portions consisting of short penetrating rhizoids. Monosporangia sessile.

Epiphytic on *Avrainvillea* along west coast of Florida, in 2-3 m.

Acrochaetium sargassi Børgesen

Taylor (1960), p. 306.

Plants to 0.7 mm; the basal cell spherical to disklike. Erect filaments with secund to opposite branching at least in upper portion, branchlets tapering toward apices; cells of main axis 5.5 μm in diameter, 9-18 μm long; cells of branchlets 2-3 μm in diameter and 30-40 μm long. Monosporangia infrequent, sessile or stalked.

On algae and *Thalassia,* in shallow waters.

Acrochaetium seriatum Børgesen

Taylor (1960), p. 310.

Plants consisting of a basal disk of filaments not penetrating host; erect filaments to 1 mm. Upper branching secund. Cells of main axes 8-10 μm in diameter and 22 μm long; cells of small branchlets 6-7 μm in diameter and 30 μm long. Monosporangia oval and sessile or stalked.

On coarse algae and sea grasses.

Acrochaetium thuretii (Bornet) Collins *et* Hervey

Taylor (1960), p. 310.

Plant with an irregular disk, 60-120 μm diameter. Erect filaments to 4 mm, cells 12 μm in diameter and to 70 μm long. Branching primarily basal, although erect branches showing some alternate-irregular branching; disk parenchymatous in appearance.

An epiphyte of sea grasses found near Anclote Key (Ballantine, 1972).

Kylinia crassipes (Børgesen) Kylin
Taylor (1960), p. 300.

Plants small to about 70 μm tall; basal cell ovoid to globular, to about 12 μm in diameter, giving rise to widely spreading filaments. Cells of filaments barrel shaped, 5-7 μm in diameter and 5-9 μm long; tips of branches may have colorless hairs.

Epiphytic on red algae (*Hypnea* and *Laurencia*) in shallow water.

Family Chaetangiaceae

Galaxaura obtusata (Ellis *et* Solander) Lamouroux
Taylor (1960), p. 342.

Plants light pink to rose red in color, 7-10 cm tall, branched and terete; branches joined at forks and only lightly calcified. Cortex of one inner layer of large cells and one to two outer layers of smaller cells. Central region composed of free filaments.

Plants on rocks in shallow water, found in the summer.

Scinaia complanata (Collins) Cotton
Taylor (1960), p. 334.

Plants pale pink to rose red, to 8 cm tall, branching dichotomous eight or nine times. Thallus soft, consisting of an obscured central medulla of six to eight filaments that give rise to radial, dichotomously branching cells, which in turn give rise to the cuboidal or rectangular epidermal cells (22-34 μm in size).

Plants found in depths of 2-10 m, growing offshore on limestone collected in spring.

ORDER GELIDIALES
Family Gelidiaceae

Gelidiella acerosa (Forsskål) Feldmann *et* Hamel
Taylor (1960), p. 351.

Plants to 12 cm tall, tough and wiry; basal portions prostrate and attached by rhizoidal outgrowths. Erect branches tapered and terete, with an apical cell at each tip. Branching pinnate to secund, branchlets similar to main axes. Medulla of thick-walled filaments and a cortical zone of cuboidal cells.

On rocks and jetties in the lower littoral and sublittoral zones.

Gelidium crinale (Turner) Lamouroux
Taylor (1960), p. 355.

Plants forming large tufts, to 5 cm high, branches about 0.5 mm wide; the wiry main axes decumbent and spreading over substratum and attached by rhizoidal holdfasts. Pinnately branched and branches primarily in upper portions. "Rhizines" (thick-walled filaments) found on the edge of medulla and the inner cortex. Tetraspores cruciate and scattered on erect branches.

Growing on rocks at lower littoral level.

Gelidium pusillum (Stackhouse) Le Jolis
Taylor (1960), p. 354.

Plants small, to 0.5 cm, branches thin and flattened at the tips, to 0.7 mm broad; branching somewhat pinnate. Rhizines most obvious in older, basal portions, and found in the edge of the medulla. Tetraspores cruciate and scattered.

Usually growing in the midlittoral zone on rocks and jetties, common in the winter along causeways in Tampa Bay.

Pterocladia americana Taylor
Fig. 53
Taylor (1960), p. 360.

Plants to 6 cm in height, main axes difficult to determine, branching sparse, irregularly to slightly pinnately branched; branchlets tending to become flattened blades, 500-600 μm wide and 75 μm thick. Basal portions decumbent, axes with rhizoidal holdfasts. Rhizines apparent in older portions, scattered among the thin-walled filaments of the medulla. Tetraspores cruciate and in rows.

Common on jetties and seawalls; appears in the upper littoral zone in the spring and the lower littoral zone in the summer. The form identified here is somewhat more distinct in its pattern of branching than typical material (see Taylor, 1960).

Family Wurdemanniaceae

Wurdemannia miniata (Draparnaud) Feldmann et Hamel
Taylor (1960), p. 361.

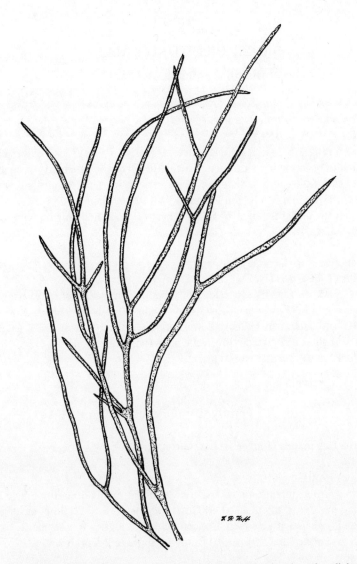

Fig. 53. *Pterocladia americana*. A general habit sketch showing the dichotomous branching of the upper portion. *x 2.5*

Plants bushy, main axes difficult to determine, texture wiry, coarse; medulla of long, thin-walled filaments. Tetraspores zonate, and borne in enlarged tips of branchlets.

Forming dull red or pink clumps on rocks in shallow water. Not common at this level, probably frequent at greater depths than 5 m. The placement of this family is open to question because of the zonate tetrasporangia.

ORDER CRYPTONEMIALES

Family Squamariaceae

Hildenbrandia prototypus Nardo
    Taylor (1960), p. 369.
    Plants orange to red, forming thin crusts 0.2-0.5 mm thick. Lower cells (hypothallus) supporting an upper (perithallus) layer of irregularly branched rows of cells, 4.0-6.5 μm in diameter. Conceptacles scattered over plant, when mature tetrasporangia 16-30 μm long and 9-14 μm in diameter.
    Frequently in depths of 2-20 m on limestone and shell. Also common in brackish water.

Peyssonnelia rubra (Greville) J. Agardh
    Taylor (1960), p. 371.
    Plants pink or red, lightly calcified, lobed and membranous; adhering to rock by rhizoids. Lower cell row (hypothallus) producing upper rows or filaments of cells (perithallus). Tetrasporangia in superficial areas (nemathecia) of thallus, associated with sterile hairs (paraphyses).
    On rocks in deep water (to 10 m).

Family Corallinaceae

Amphiroa fragilissima (Linnaeus) Lamouroux
    Fig. 54
    Taylor (1960), p. 403.
    Plants having dichotomous to trichotomous branching, the angle of branching narrow; segments 150-600 μm in diameter, 8 to 20 times as long as broad, ends of segments swollen. Conceptacles lateral, few to a segment.
    On coarse algae (like *Sargassum*) found offshore in 2-5 m of water.

Amphiroa rigida Lamouroux var. **antillana** Børgesen
    Taylor (1960), p. 404.
    Plants having irregular to unequal dichotomous branches; upper ends of segments forking to form dichotomous branches. Older, lower segments 1.0-1.5 mm in diameter and found growing on limestone in 8-10 m of water.

Corallina cubensis (Montagne) Kützing
    Taylor (1960), p. 409.
    Plants small, 1-3 cm tall, densely branched, branching opposite or some-

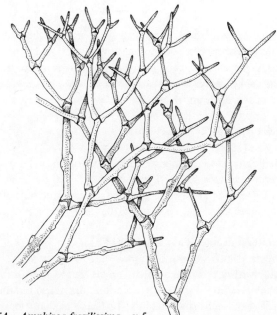

Fig. 54. *Amphiroa fragilissima.* x 5

what pinnate to verticellate. Segments of main axes 180 µm in diameter and 600 µm long; segments of branchlets 100 µm in diameter and tapering to apex. Conceptacles lateral, replacing branchlet.

Plants forming clumps, on rock and limestone in 1-5 m of water, usually offshore.

Fosliella atlantica (Foslie) Taylor
Fig. 55
Taylor (1960), p. 387.

Plant a white crust, consisting of a single layer of cells, cells 10-15 µm in diameter and 18-30 µm long; conceptacles conical, 500-800 µm in diameter, and with a single pore.

Common on *Thalassia* blades, the entire plant reaching 5 mm to 1 cm in diameter. Conceptacles of the genus *Fosliella* are in part identified by having a single pore. The genus *Melobesia* has conceptacles with numerous pores. The latter is common in the Florida Keys but has not been found along the west coast of Florida.

Fosliella farinosa (Lamouroux) Howe
Taylor (1960), p. 388.

Plant a crust, one to three cells thick; surface cells variable in size 12-20 μm by 7-19 μm, some becoming swollen to bulbous and tapering to a hair (trichocytes). In cross section the lower cells 15-18 μm broad and 18-30 μm long. Cystocarpic conceptacles 140-250 μm in diameter in surface view and having a single pore.

On coarse algae and *Thalassia* blades, common in grass beds.

Fosliella lejolisii (Rosanoff) Howe
   Taylor (1960), p. 387.
   Thallus crustose, the edge of the crust of one cell layer, three cells thick in the center; surface cells 6-7 μm wide and 6-10 μm long. Cystocarpic conceptacles 150-300 μm in diameter in surface view and having a single pore.

On coarse algae and *Thalassia* blades and common in shallow water.

Goniolithon decutescens (Heydrich) Foslie
   Taylor (1960), p. 398.
   Plant with a crustose base and erect, free to anastomosing terete branches; branches 5-10 mm long and to 1 mm in diameter. In cross section the crust a lower layer of cells (hypothallus) 12-15 μm in diameter and an upper layer of cells (perithallus) 8-15 μm in diameter; large, clear heterocysts with thickened walls are scattered through the tissue. Conceptacles 300-400 μm in diameter.

Plant on rocky-limestone surfaces and also found growing near mangrove roots in clear water.

Jania adhaerens Lamouroux
   Taylor (1960), as *J. adherens*, p. 413.
   Plant 1 to 3 cm tall and consisting of dichotomously branched segments, angle of branching 45° to 60°; lower segments 100-200 μm in diameter and

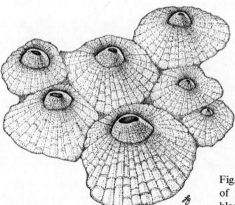

Fig. 55. *Fosliella atlantica*. Habit sketch of plant as seen when epiphytic on blades of *Thalassia. x 45*

two to four diameters long; basal segments showing slight bifurcation for branching; apices conical. Conceptacles 300-340 μm long.

Growing in clumps on rocks and limestone to depths of 5 m.

Jania capillacea Harvey
Fig. 56

Taylor (1960), p. 412.

Plants with wide dichotomous branches; 4-10 mm tall or taller; segments 45-100 μm in diameter and four to ten diameters long; conceptacles usually in between branch pairs, arising from a swollen basal segment.

Common on coarse algae (like *Sargassum*) in 2-5 m of water.

Jania pumila Lamouroux
Taylor (1960), p. 414.

Plants to 3 mm tall from a crustose base branching sparse with two to three dichotomies; segments stout, 60-150 μm in diameter and 200 μm long. Conceptacles arising from swollen segments, which also give rise to branches.

The plant is common on coarse algae.

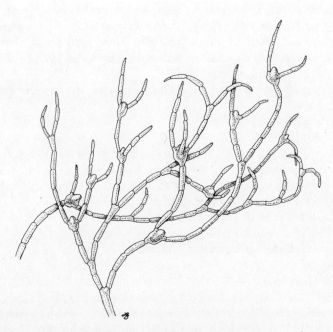

Fig. 56. *Jania capillacea*. Conceptacles are visible at branch junctions. *x 10*

Lithothamnium occidentale (Foslie) Foslie
Taylor (1960), p. 385.
Young plants having unattached lobelike branches; older forms densely branched; the branches growing together, only the tips projecting; cells of perithallus (outer cells) 6-10 μm in diameter and 10-32 μm long.
On limestone offshore in 5-20 m of water.

Lithothamnium syntrophicum Foslie
Taylor (1960), p. 381.
Plants 2-3 cm in diameter, crustose and conforming to the substratum; not always adherent; new layers growing over previous ones, each layer 200-300 μm thick; the surface with nodules; cells of upper layers 4-7 μm in diameter; conceptacles large, 400-600 μm in diameter.
On limestone in deeper water (from 5 m).

Family Grateloupiaceae

Grateloupia filicina (Wulfen) C. Agardh
Taylor (1960), p. 424.
Plants olive gray to rose red; elongated or strap-shaped fronds; the fronds to 75 cm (usually 10 cm) long, 2-5 mm wide, and tapering from base to apex; usually much branched from the margins. Some proliferation from the blade surface. Branchlets flattened to filiform (2 mm wide). Medulla of loose anastomosing filaments.
Commonly in the lower littoral zone, on rocks, plant much varied in form.

Halymenia agardhii De Toni
Taylor (1960), p. 417.
Plants rose red, bushy, 1-2 dm tall, attached by a discoid holdfast; branching dichotomous, 4 to 12 times; branches 4-7 mm in width, terete except below forkings where slightly flattened. Medulla of slender, loosely arranged filaments. Cortex of large inner and smaller outer cells.
Collected in 2-10 m of water off the west coast.

Halymenia floresia (Clemente) C. Agardh
Fig. 57
Taylor (1960), p. 418.
Plants to 3 dm tall, having a gelatinous texture and profusely branched; consisting of a flattened main axis or axes, with marginal, pinnately arranged branchlets. Main axes to 20 mm wide and 2 mm thick. Branchlets attenuate to terete at the apices. Medulla of loose filaments, with occasional stellate ganglia (star-shaped cells giving rise to filaments at all points).

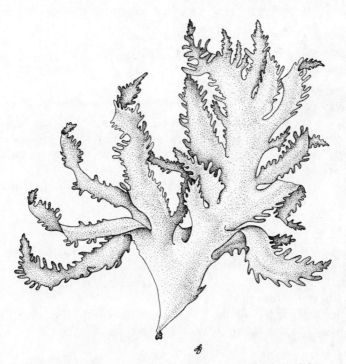

Fig. 57. *Halymenia floresia.* A young plant with but one main axis. *x 0.5*

Found in drift and reported in water of 2-10 m. Probably a deepwater species.

Halymenia gelinaria Collins *et* Howe
   Taylor (1960), p. 420.
   Plants deep purple in color; blades gelatinous to fleshy in texture, blades to 50 cm broad; plants sparsely branched at base, blades orbicular with lobed to dentate margins. Medulla having occasional stellate ganglia to 50 μm in diameter. Subcortex a close network of anastomosing filaments.
   One specimen collected attached, in 10 m off Tampa Bay.

Halymenia pseudofloresia Collins *et* Howe
   Taylor (1960), p. 418.
   Plants to 4 dm long, 2.5 dm wide, gelatinous. Main axes bearing numerous marginal, linear branchlets. Main axes to 2 cm wide, more slender. Plants bushy. Medulla of loose anastomosing filaments.
   Collected in drift, and attached in 10 m of water.
   Note: Recent collections from the Florida "middlegrounds," an area 70 to 100 miles off the west coast of Florida, have yielded a number of specimens that may be *H. floridana* J. Agardh.

Family Kallymeniaceae

Kallymenia limminghii Montagne
  Taylor (1960), p. 432.
  Plants with one or more orbicular blades, blades 0.5-15 mm in diameter,
arising from a slender stipe. The blades proliferating, margins erose-dentate to
strongly crenulate. In cross section a medulla of strongly interwined and
anastomosing filaments is visible.
  Collected as lithophytes and epiphytes offshore Anclote Key in 10 m of
water.

Kallymenia perforata J. Agardh
  Blades irregular to orbicular in shape, to 1 dm in diameter; blades having
numerous holes, (1-3 cm diameter); in cross section the blades with a cortex
of small pigmented cells, a subcortex of large, clear cells, and a loose medulla
of widely spaced filaments.
  A large population of this species was found in 9 m of water offshore
Anclote Key in December; all material was sterile.

ORDER GIGARTINALES

Family Gracilariaceae

Gracilaria armata (C. Agardh) J. Agardh
  Plants 1-5 dm, cartilaginous in texture, branches terete, 3-5 mm in diameter. Main axes radially and abundantly branched, upper branches subdichotomous to secund. Three-layered cortex, cortical cells 7-11 μm in diameter; thickened central cells 210-260 μm in diameter. Deep red to brown in
color.
  Found free-floating in bays in the autumn and winter.

Gracilaria blodgettii Harvey
  Taylor (1960), p. 449.
  Plants 1-3 dm, alternately branched and spreading. Plants deep brown or
greenish brown. Branches constricted at the base and tapering to the tips, to
1.8 mm in diameter. Main axis 1.5-3.0 mm in diameter. Cortex of three
layers, outer cells 15-20 μm in diameter, inner cells (two layers) 30-125 μm in
diameter. Large central cells thin walled, 650-970 μm in diameter. Cystocarps
150-175 μm in diameter.
  Attached to rock and limestone in upper sublittoral region in areas of
moderate salinity (20 0/00).

Gracilaria cervicornis (Turner) J. Agardh
Taylor (1960), p. 445.
Plants to 2.5 dm tall, fleshy-membranous in texture. Pinnately branched, the upper branches terete and subdichotomous. Older parts flattened. Upper branches often with dentate margins. Main axes 2-5 mm broad or wider. Pale red to dark brown in color.
On rocks in areas of moderate wave action.

Gracilaria compressa (C. Agardh) Greville
Taylor (1960), p. 444.
Plants to 2.5 dm tall, dichotomously branched below, the upper branches radially alternate to secund, branching giving aspect of a cluster of plants. Each fork compressed and with distal terete portions. Central cells thin walled, 200-310 μm in diameter. Cortex three-layered, outer cells 8-9 μm in diameter; inner two layers of cells 9-25 μm in diameter. Deep brown to purple in color.
Plants loose on bottom in bays and areas of quiet water.

Gracilaria cylindrica Børgesen
Taylor (1960), p. 450.
Plants erect, 2-3 dm tall, fleshy; usually a deep red color but becoming bleached in clear waters. Sparsely branched, branches terete throughout. Major axis short, branching to form a number of secondary axes. Branchlets constricted at forkings, branches to 4 mm in diameter, more commonly 2-3 mm; in cross section consisting of a central region of parenchymatous cells 400-500 μm in diameter, a subcortex of one to two cell layers of smaller cells and two layers of pigmented cortical cells.
Collected adrift and attached by diving and dredging in 10-40 m of water offshore.

Gracilaria damaecornis J. Agardh
Taylor (1960), p. 443.
Plants 7-15 cm tall, cartilaginous in texture. Branching irregular above, with upper short branches crowded, generally blunt, sometimes secund. Dichotomously branched below, the branches 1.5-4.0 mm in diameter, terete throughout.
Uncommon, attached to rocks offshore.

Gracilaria debilis (Forsskål) Børgesen
Taylor (1960), p. 442.
Large (to 2 dm), cartilaginous plants, usually red to yellow brown, with secund or recurved branches developing from a prominent main axis, terete throughout. Central region composed of thick-walled cells, 200-360 μm in

diameter. Outer cortical cells 9-13 μm in size, inner cells to 70 μm (35-70 μm).

Plants may occur as dense mats (1-2 dm in diameter) or as erect forms.

Gracilaria foliifera (Forsskål) Børgesen
Fig. 58
Taylor (1960), p. 446.

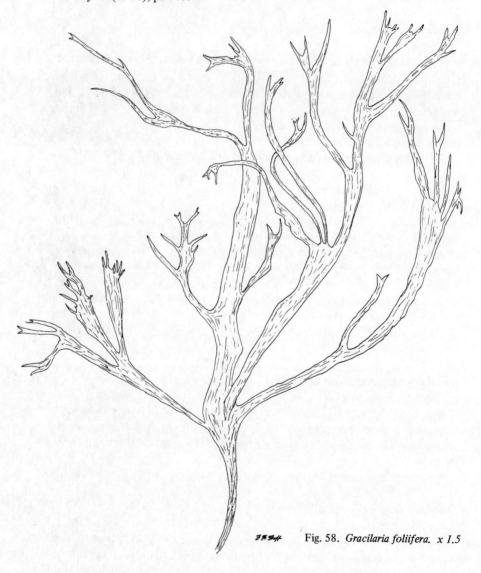

Fig. 58. *Gracilaria foliifera.* x 1.5

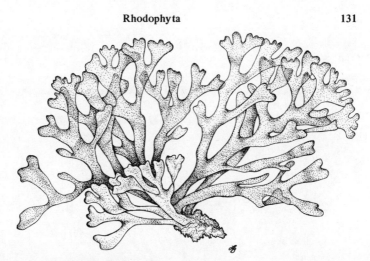

Fig. 59. *Gracilaria mammillaris.* x 0.5

Plants flat throughout, dichotomously or alternately branched above. Strap-shaped blades often proliferous, especially near the base; blades to 18 mm broad. Plants 5-30 cm tall. Cortex of six to eight layers of small cells. Green to dark purple or brown in color.

Occurs in sea grass beds and on rocks and rock jetties.

Var. **angustissima** (Harvey) Taylor

Taylor (1960), p. 447.

Branches terete to compressed below, flattened above. Subdichotomous to irregularly branched, and with few proliferations. Plants 5-25 cm tall. Forkings at narrow angles.

On rock jetties; plants with much morphological variation.

Gracilaria mammillaris (Montagne) Howe

Fig. 59

Taylor (1960), p. 447.

Plants 0.5-1.5 dm tall, orange to dull red in color. Branching dichotomous with wide forkings; having a short stalk. Fronds flattened 3-9 mm broad, oblong or linear in shape, with a firm texture. Thick-walled central cells 50-125 μm in diameter. Upper branches sometimes irregularly curled.

Plants of shallow sea grass beds are bleached and firm in texture; plants of deeper water are rosy in color and less cartilaginous in texture. Plants are frequently loose on bottom, showing vegetative growth by fragmentation.

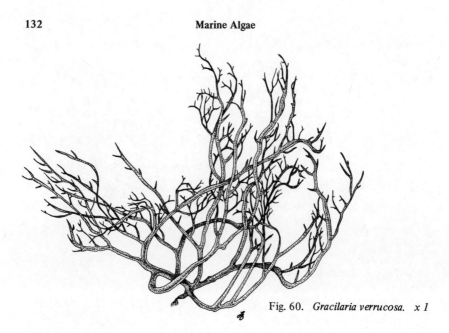

Fig. 60. *Gracilaria verrucosa. x 1*

Gracilaria verrucosa (Hudson) Papenfuss
Fig. 60
Taylor (1960), p. 441.

Thalli 1-5 dm tall, firm, fleshy. Fronds 0.5-3.0 mm in diameter, irregularly or alternately branched, tapering at both ends. Central cells thin walled, 180-300 (to 480) μm in diameter. Cortex of three to four cell layers, the outermost layer of cells 5-12 μm in diameter, the inner cells 20-30 μm in diameter; numerous hair cells are present on the surface. Tetrasporangia 30-33 μm long. Cystocarps to 230 μm in diameter. Spermatia in pits 45-75 μm deep.

In grass beds, attached to shell, probably the most common form of *Gracilaria*. Frequently covering the bottom in drift with *Spyridia*.

Gracilariopsis sjoestedtii (Kylin) Dawson
Taylor (1960), as *Gracilaria sjoestedtii* Kylin, p. 449.

Thalli 20-50 cm tall, fronds 1.5-3.0 mm in diameter. Plants loosely and irregularly branched, with many long, cylindrical, primary branches. Central cells 350-420 μm in diameter, grading to a six-layered cortex of cells 9-12 μm in diameter. Tetrasporangia 36-38 μm long, in sori. Spermatia in a continuous layer, often spreading over the entire cortical surface. Cystocarps low, elliptical, 0.8-1.8 mm in diameter.

Infrequent, plants found attached in grass beds.

Family Rhabdoniaceae

Catenella repens (Lightfoot) Batters
  Fig. 61
  Taylor (1960), p. 462.

Plants prostrate, consisting of oval to spindle-shaped segments, measuring about 0.2-0.5 cm in diameter. Segments having a medulla of filaments and a small central filament (uniaxial); plant developing from a single apical cell. Branching dichotomous to trichotomous, with rhizoidal outgrowths at points of forking. Tetraspores scattered over segments.

Plants pale to dark red and appear as minute cacti because of the segmented structure. Found on prop roots of the red mangrove, with *Bostrychia*.

Family Solieriaceae

Agardhiella ramosissima (Harvey) Kylin
  Taylor (1960), p. 457.
  Bushy plants, 20-30 cm in height. Fleshy to gelatinous in texture, yellow

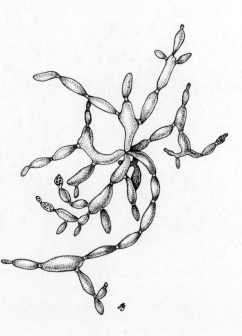

Fig. 61. *Catenella repens*. Tetrasporangia are shown on some of the ultimate branchlets. *x 5*

brown to dull red in color. Axis opposite to irregular but distichously branched. The larger portions are distinctly compressed.

A winter-spring plant attached on rocks along causeways in Tampa Bay and offshore to depths of 30 m. The plant can be confused with *Eucheuma acanthocladum,* which also has a loose medulla. Wynne and Taylor (1973) have proposed that this species should be renamed *Neoagardhiella ramosissima* (Harvey *ex* Kützing) Wynne and Taylor.

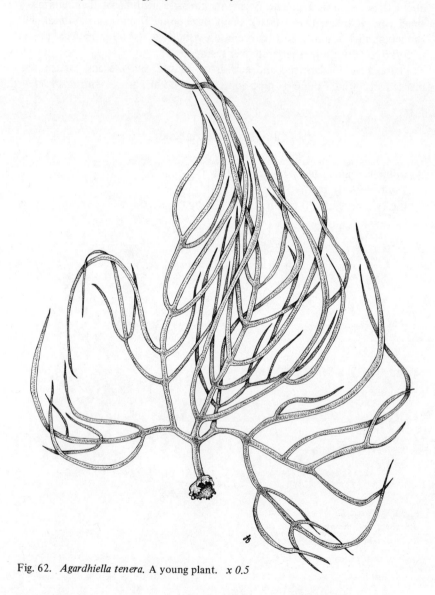

Fig. 62. *Agardhiella tenera.* A young plant. x 0.5

Agardhiella tenera (J. Agardh) Schmitz
Fig. 62
Taylor (1960), p. 456.
Bushy plants to 3 dm tall, with a fibrous base, radially branched and terete, firm to fleshy in texture; branches 1-4 mm in diameter. Branchlets taper toward base and tip, and are slightly constricted at the base. Medullary filaments thick walled and continuous in longitudinal section. Tetrasporangia 27 to 33 μm long, 17 to 23 μm in diameter.

A common spring-summer plant in Tampa Bay and along the west coast, attached to rocks, shells, and pilings. Wynne and Taylor (1973) have renamed this species *Solieria tenera* (J. Agardh) Wynne and Taylor. In the same paper, a third species, *Agardhiella baileyi* (Harvey *ex* Kützing) Taylor is renamed *Neoagardhiella baileyi* (Harvey *ex* Kützing) Wynne and Taylor. *Agardhiella baileyi* is described as a cold-water plant, occurring northward along the east coast to southern New England. *Agardhiella baileyi* lacks thick-walled, continuous, medullary filaments and has a simple basal attachment. In contrast, *A. tenera* (*Solieria tenera*) is described as a warm-water plant occurring in the Gulf of Mexico, with thick-walled, continuous, longitudinal medullary filaments and a loose, fibrous basal attachment.

Eucheuma acanthocladum (Harvey) J. Agardh
Taylor (1960), p. 458.
Plants to 3 dm or larger, more commonly about 1 dm. Main branches widely spread, slightly flattened to oval; the branching of the main axis occurring in a zigzag pattern and primarily in one plane. The branches irregularly alternate to somewhat dichotomous. Texture firm to somewhat slimy; color pale red to brown red. In cross section, the medulla consisting of loose filaments, like *Agardhiella*.

Collected by diving and dredge in offshore stations (10 to 40 m); not common. Plant may be confused with *E. gelidium* (J. Agardh) J. Agardh, which has a more dense, congested ultimate branching and broader, more flattened main axis. The species *E. gelidium* is common in the Florida Keys. The taxonomic status of *E. acanthocladum* is open to question.

Eucheuma isiforme (C. Agardh) J. Agardh
Fig. 63A
Taylor (1960), p. 459.
Plants erect, 3-5 dm tall, or taller; texture firm to cartilaginous, color deep red to light yellow. Branching is sparse, having two to three orders of branches; branches having opposite or whorled spines or nodules; plant developing from a main holdfast, usually erect, with few or no secondary holdfasts. In cross section the medulla composed of densely packed filaments.

Collected attached and in drift offshore, typical of deep (30+ m) water.

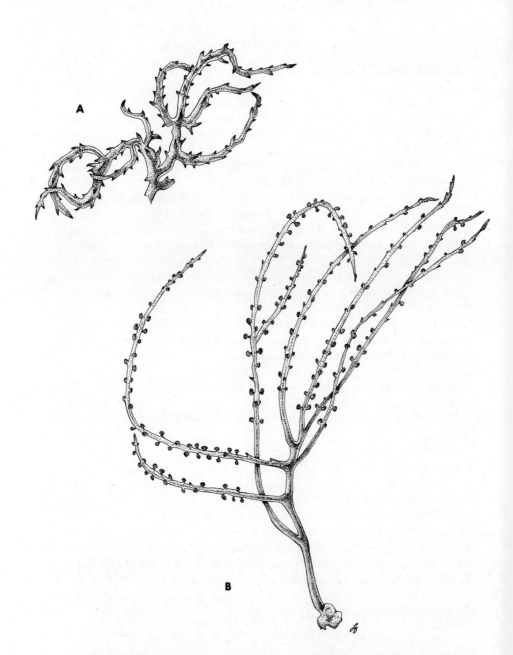

Fig. 63. A, *Eucheuma isiforme*. The species is coarse with whorls of spines or knobs and is found in shallow water (2-10 m) in the Florida Keys and in deep water (30 m+) off the west coast of Florida. *x 0.5*; B, *Eucheuma nudum*. A slender species lacking numerous spines, the occasional spines not in whorls, and having a procumbent habit. A cystocarpic plant. *x 0.5*; C (opposite), *Eucheuma nudum*. A sterile, young plant. Note the regeneration collars. *x 1*

Eucheuma nudum J. Agardh

Figs. 63B, C

As *E. acanthocladum*, Dawes and van Breedveld (1969), and *E. isiforme*, Dawes (1967), p. 66.

Plants 3-5 dm tall or taller in waters of mid depth (5-30 m). Branching from a single main holdfast; with numerous secondary holdfasts. Plant usually procumbent in growth, although it may be erect. Main axes (four to five) devoid of whorls of spines or nodules, although occasional spines can be present. Plants deep red to brown in color. Medulla, in cross section, composed of densely packed filaments. Cystocarps on short stalks; tetrasporangia zonate, in the outer cortex; antheridia scattered among the outer cortical cells.

A common offshore plant of year-round occurrence is identified as *E. nudum* on the basis of collections of similar material from the east coast of Florida. Plants found in shallow water (9 m) demonstrate a complete life

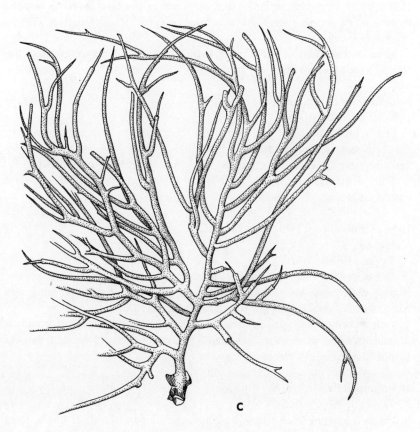

C

history with new sporelings developing in the late winter months (February to March). Plants larger than 1 m in length are common in 30 m of water.

Meristotheca floridana Kylin
    Taylor (1960), p. 460.
    Plants foliaceous to 3.5 dm tall and 4 cm wide; fronds lobed or irregularly cleft; segments oblong and 0.5-30 cm broad. Margins dentate; surface usually smooth, but it may have dentate proliferations.
    One specimen collected in 10 m of water offshore.

Family Hypneaceae

Hypnea cervicornis J. Agardh
    Taylor (1960), p. 466.
    Plants 2-15 cm tall, forming tangled and extensive tufts; main axis subdichotomously branched, alternate and cervicorn at the tips. Branches smooth or with a few simple spines. Plants in loose, soft cushions, pale red to brownish purple in color.
    Described as plant of heavier surf than *H. spinella*. A major feature is the lack of a well-defined main axis.

Hypnea cornuta (Lamouroux) J. Agardh
    Taylor (1960), p. 467.
    Plants bushy to 2 dm tall, alternately branched, the branches tapering to the tips and bearing numerous spiny branchlets to 2.5 mm long. Older branches bearing acute stellate-tipped branchlets with three to six points.
    The plants resembling *H. spinella* or *H. cervicornis*, but the distinct cornute spines are obvious, common in mangrove swamps.

Hypnea musciformis (Wulfen) Lamouroux
    Fig. 64
    Taylor (1960), p. 467.
    Plants bushy 10-30 cm tall. Main branches about 1-2 mm in diameter, bearing many spur branchlets, incurved in the upper branches; branch tips bare, typically swollen and strongly hooked (crosier tip).
    Dull brown or red in color, sometimes green or black. A common epiphyte in sea grass beds, where it is often unattached or entangled. Attached, densely branched plants are common on exposed rocks.

Hypnea spinella (C. Agardh) Kützing
    Taylor (1960), p. 465.
    Plants branching in all planes, branches smooth or with simple spines. Main

Fig. 64. *Hypnea musciformis.* Crozier-hooked branches are shown on this young, laxly branched form collected on rock jetties. *x 1*

axes apparent and mode of branching visible, usually heavily spinose. The upper segments to 2.2 mm in length, tapering or spinelike. Plants subcartilaginous in texture, dull red in color; usually in dense mats or tufts to 2.5 cm in thickness.

Loose, free-floating or entangled among other algae.

ORDER RHODYMENIALES
Family Champiaceae

Champia parvula (C. Agardh) Harvey
Fig. 65
Taylor (1960), p. 490.
Plants tufted and loose, pale red, brown, or green in color, sometimes almost colorless, branching alternate. Membranous in texture, the hollow

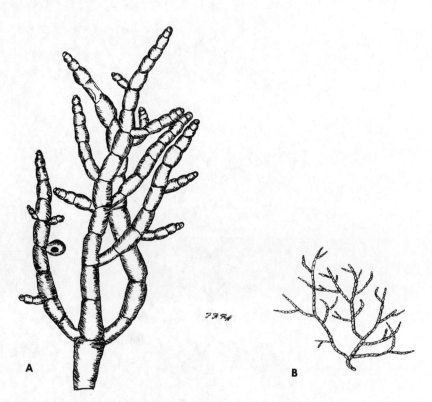

Fig. 65. *Champia parvula*. A, *x 2.5* B, *x 0.5*

barrel-shaped segments are visible. Segments separated by multicellular septae. Branches 1.5 mm in diameter or less.

Common on sea grasses and other algae in shallow water.

Lomentaria baileyana (Harvey) Farlow
Taylor (1960), p. 487.
Tufted plants, 3-7 cm tall, densely branched and soft in texture; branching alternate, though irregular. Plants hollow, terete and tapered toward the ends. Upper segments 1-2 cm long and 0.5-1.5 mm in diameter. Dull purple to pink.

Common on sea grasses and coarse algae in shallow water.

Lomentaria rawitscheri Joly
Taylor (1960), p. 487.
Plants tufted 5-7 cm tall. Branching irregular and decumbent, bearing many erect simple branchlets of 1-2 mm. Branchlets oval with a pointed tip. Brown to red in color.

Found on rocks in the surf zone at Mullet Key, old blockhouse.

Family Rhodymeniaceae

Botryocladia occidentalis (Børgesen) Kylin
Fig. 66
Taylor (1960), p. 483.
Plants to 2.5 dm tall with loose, slender main axes beset with spherical to obovoid branchlets (bladders). Bladders are 4-5 mm long, pedicellate, and very numerous along the branches. Pale rose to yellow brown in color.
Plant attached to limestone, from depths of 5-20 m.

Chrysymenia enteromorpha Harvey
Taylor (1960), p. 479.
Plants 20-25 cm tall, terete or slightly compressed, hollow and delicately membranous. Thallus divided into cylindrical segments by constrictions. Medullary filaments absent.
Occasional plant in drift, probably growing in deeper water.

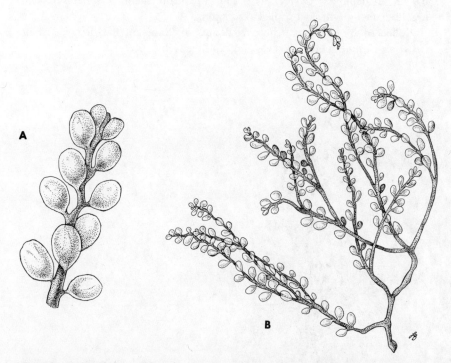

Fig. 66. *Botryocladia occidentalis.* An enlarged view of the arrangement of bladders is shown in A. A, *x 2.5* B, *x 1*

Fauchea hassleri Howe *et* Taylor
    Fig. 67
    Taylor (1960), p. 476.
    Plants dichotomously branched, 5-10 cm tall, main axis 3-5 mm in diameter, to 1 cm broad; final dichotomies rounded, 0.4-1.0 mm broad. Blades flattened, radiating in a dichotomous fan; medulla of two to three layers of polygonal cells, 110-150 μm in diameter, cortex of two to three cells, 6-9 μm in diameter.
    Collected by dredge offshore in depths of 20-60 m (Dawes and van Breedveld, 1969).

Fauchea peltata Taylor
    Taylor (1960), p. 477.
    Plants consisting of a short (1.0-1.5 cm) stipe of about 4-6 mm thickness, and a thin, soft peltate blade 3.5-5.0 cm in diameter; plants usually incrusted with bryozoans, central cells of blade hexagonal and clearly visible through cortex of two cell layers. Central cells 300-400 μm in diameter, cortex cells 6-9 μm in diameter. Cystocarps found as pendant, marginal spheres 1.4-2.0 mm in diameter with an ostiole 0.5 mm long.
    Collected by dredge offshore in depths of 20-60 m; the first report of cystocarpic reproduction (Dawes and van Breedveld, 1969).

Fig. 67. *Fauchea hassleri. x 1.25*

ORDER CERAMIALES

Family Ceramiaceae

Callithamnion byssoides Arnott *ex* Hooker
Taylor (1960), p. 506.

Plants in tufts, 3-8 cm high, light pink, filaments ecorticate; cells of main filaments 40-100 μm in diameter; the ultimate branchlets slender, 10-20 μm in diameter, the cells 6 to 10 diameters long; the axes repeatedly branched; tetrasporangia sessile.

On sea grasses in 2-5 m of water, offshore.

Callithamnion cordatum Børgesen
Taylor (1960), p. 507.

Plants small, 2-4 cm tall, main axes abundantly branched above, filaments ecorticate or with a few slender rhizoidal filaments; main axes 160-200 μm in diameter and cells 300 μm long; cells of branchlets 8 μm in diameter; branching alternate, tetrasporangia sessile.

On coarse algae, uncommon.

Callithamnion roseum (Roth) Harvey
Taylor (1960), p. 508.

Plants 4-7 cm tall, branching alternate and radial; the cells 124-350 μm in diameter and 3 to 4 diameters long. The older portions becoming corticated by rhizoids arising from base of branches; the branchlets 38-42 μm in diameter and the cells to 24 diameters long. Tetrasporangia nearly spherical, 45-70 μm in diameter.

Coating stones and *Thalassia* blades.

Centroceras clavulatum (C. Agardh) Montagne
Fig. 68
Taylor (1960), p. 537.

Plants dark red and forming dense tufts 10-20 cm tall; filaments 50-200 μm in diameter, the axes completely corticated, the internodes to 0.75 mm long; the apices strongly incurved; a row of spines at each node, the spines consisting of two hyaline cells.

Common on rocks and the base of mangrove roots; forming a definite band in the midlittoral zone in both exposed and protected areas.

Ceramium byssoideum Harvey
Fig. 69
Taylor (1960), p. 528.

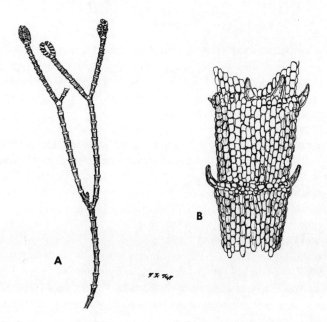

Fig. 68. *Centroceras clavulatum*. A view of the arrangement of spines at the nodes is shown in B. A, *x 0.5* B, *x 10*

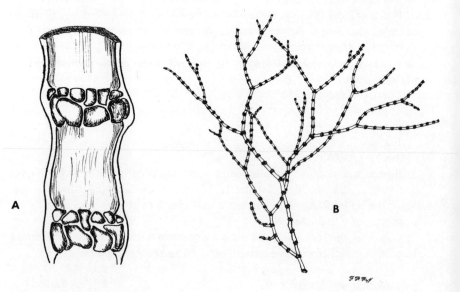

Fig. 69. *Ceramium byssoideum*. The pattern of cortication is shown in the left sketch. A, *x 0.5* B, *x 10*

Plants dichotomously branched, branches 60-90 μm in diameter; nodal cortication varying from 4 to 8 to 10 rows of cells, the lower rows having rectangular-shaped cells with their long edge transversely arranged; tetrasporangia one to two at each node and immersed within nodal cortication, tetrasporangia to 60 μm in diameter.

A more elongated form has been described with alternate branching and fewer (three to four) bands of cells (Ballantine, 1972). Plants occurring on sea grasses and coarse algae.

Ceramium codii (Richards) Feldmann-Mazoyer
Taylor (1960), p. 526.

Plants minute, about 1 cm tall, with prostrate filaments 30-50 μm in diameter; simple, erect filaments arising at every node; each node consisting of about two rows of cortical cells, irregularly arranged; tetrasporangia 20-30 μm in diameter to 45 μm long, half-covered by cortical cells.

Epiphytic on *Codium*, collected at Mullet Key.

Ceramium corniculatum Montagne
Taylor (1960), p. 530.

Plants with upper axes covered by nodal cortications; on the lower segments the cortical cells are elongated and parallel to the filament axes; the internodes hyaline and three to four diameters long. Tetrasporangia verticellate in central parts of the nodes.

Collected occasionally as epiphyte on *Thalassia*.

Ceramium fastigiatum (Roth) Harvey
Taylor (1960), p. 526

Plants rose red, basal portions creeping, branching dichotomous; cortication of mature nodes consisting of four to six cell rows; cells 60-150 μm in diameter and 55-65 μm long. The internodes 75-150 μm in diameter, 600-1,500 μm long. Tetrasporangia (33-65 μm in diameter) partially covered by nodal cells.

On sea grasses and coarse algae, and also floating as a mat.

F. **flaccida** Petersen
Taylor (1960), p. 527.

Plants in tufts to 7 cm; nodal cells in two rows, the upper row consisting of smaller cells and the lower row of larger cells; nodes 50 μm in diameter, 20-40 μm long; internodal cells 50 μm in diameter, to 400 μm long.

Common on mangrove roots and oyster shells, in brackish water.

Ceramium floridanum J. Agardh
Taylor (1960), p. 534.

Plants rose red, to 7 cm tall; the upper portions showing a regular, alternate to subpinnate branching pattern; ultimate branchlets 30-60 μm in diameter; the older portions irregularly branched; nodes to 300 μm in diameter and 400 μm long. Tetrasporangia whorled in the smaller branches, to 75 μm in diameter.

Uncommon, found growing on rocks and such coarse algae as *Gracilaria* and *Grateloupia.*

Ceramium strictum (Kützing) Harvey
Taylor (1960), p. 530.

Plants 2-8 cm tall, branching dichotomous and spreading; nodes 210-340 μm in diameter and 85-270 μm long. Cortical cells angular and of varying size; internodal cells to 280 μm in diameter and 2,500 μm long. Tetrasporangia covered by nodal band or slightly exposed 50-60 μm in diameter.

On sea grasses and oyster shells.

Ceramium subtile J. Agardh
Taylor (1960), p. 527.

Plants forming soft tufts, to 15 cm tall, branching dichotomous; internodes 120 μm in diameter, 10 diameters long; nodes 120 μm in diameter, consisting of three series of cells in regular rows; the middle row consists of large, subspherical cells whereas the other rows are of small cells. Tetrasporangia to 60 μm in diameter, half-covered by cortical cells.

Growing in shallow water on firm substrate and coarse algae.

Crouania attenuata (Bonnemaison) J. Agardh
Taylor (1960), p. 495.

Plants 1-3 cm tall, gelatinous, red, often whitened with light calcification; branching irregularly alternate; branchlets determinate to 0.2-5 mm diameter; cells of axes 170 μm in diameter and 300 μm long. Ecorticate.

On rocks, shells in shallow water.

Griffithsia globulifera Harvey
Fig. 70

Taylor (1960), p. 517.

Plants erect to 6 cm, alternately branched; cells 0.6-3.2 mm in diameter, swollen and pyriform; branches appearing moniliform; ecorticate or with a few rhizoidal filaments arising from the lower branches, branches bearing delicate, colorless branched hairs near their apices.

Epiphytic on coarse algae.

Griffithsia tenuis C. Agardh
Taylor (1960), p. 516.

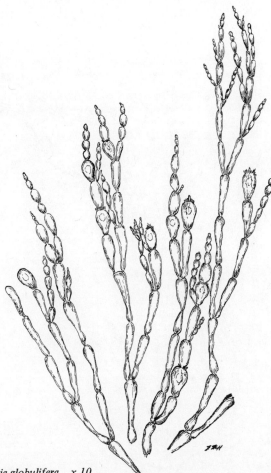

Fig. 70. *Griffithsia globulifera.* x 10

Plants erect to 7 cm tall, attached by creeping basal filaments; alternately branched; cells of main axes cylindrical 120-300 μm in diameter and three to six diameters long; apical cells shorter, with whorls of trichoblasts; tetrasporangia verticellate.

Epiphytic on coarse algae.

Seirospora occidentalis Bϕrgesen

Taylor (1960), p. 510.

Plants 1-2 cm tall, attached by rhizoids from lower axial cells; branching alternate to opposite; cells of main axes ecorticate, 85 μm in diameter, 3 to 5 diameters long; cells of branchlets 8-11 μm in diameter and 2 to 10 diameters long; sporangia sessile; thick-walled spores in branched rows of three to five subspherical cells, (seirospores), seirospores 18-20 μm in diameter.

Epiphytic on coarse algae, uncommon and apparently a winter species.

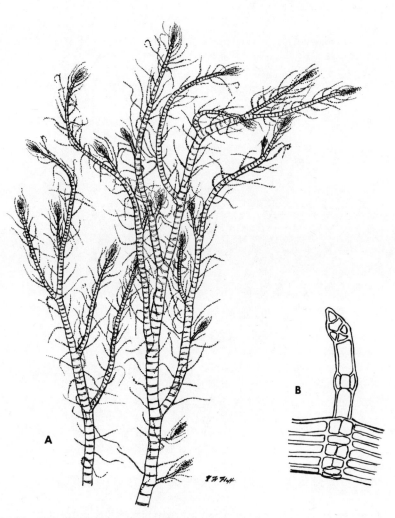

Fig. 71. *Spyridia filamentosa*. An enlargement of a single monosiphonous branch with a single spine at the tip is shown on the right. A, *x 2.5* B, *x 10*

Spermothamnion turneri (Mertens) Areschoug var. **variabile** (C. Agardh) Ardissone
Taylor (1960), p. 519.
Filaments in thick, erect tufts (1-3 cm tall) arising from a creeping base; erect axis mostly alternately branched (in variety); cells of creeping filaments 30-45 μm in diameter, cells of erect filaments 20-70 μm in diameter.
Collected as epiphytes on *Codium* in 9 m off Anclote Key.

Spyridia filamentosa (Wulfen) Harvey
Fig. 71

Taylor (1960), p. 539.

Plants deep red to straw colored, bushy, starting from a rhizoidal discoid disk holdfast, later becoming free; 1.5-2.0 cm tall; branching alternate, branches covered with short deciduous branchlets; the main axes a single row of cells, heavily corticated, the cross walls visible through the cortication; branches pointed or ending in hook-shaped tips (hemate branches). Diameter of main axes 1-2 mm; branchlets radially arranged, cells 20-45 μm in diameter and two to four diameters long; the branchlets corticated at nodes, a single spine arising from terminal pointed cell of branchlet.

A common plant of both exposed and quiet waters. In sheltered waters, it grows with *Gracilaria verrucosa* and it forms a dense bottom mat in the winter-spring.

Family Delesseriaceae

Caloglossa leprieurii (Montagne) J. Agardh
Fig. 72
Taylor (1960), p. 544.

Plants flattened throughout, composed of dichotomously branched segments about 2 mm broad, constricted at the forks. Segments 4-6 mm long, bearing a midrib. Rose red to purple in color.

Fig. 72. *Caloglossa leprieurii.* x 5

Var. **hookeri** (Harvey) Post

A common plant found attached on coquina rock at Point of Rocks
(Eiseman, in preparation), and on mangrove prop roots and pneumatophores
in Tampa Bay. Branching is irregular with fronds in clusters.

Cryptopleura fimbriata (Greville) Kützing
Taylor (1960), p. 554.

Plants 7-12 cm tall, thickened, blades of irregular width, lobed above.
Upper segments oblong, obtuse. Sori mainly in marginal leaflets. Deep red in
color.

Epiphytic on mangrove roots.

Grinnellia americana (C. Agardh) Harvey
Fig. 73
Taylor (1960), p. 547.

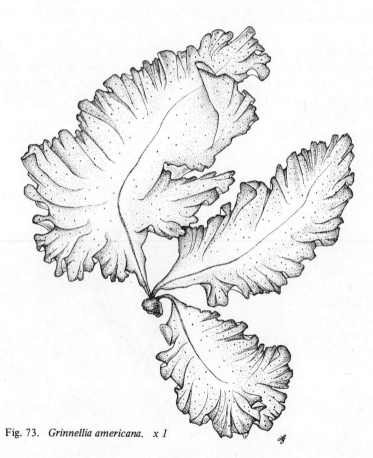

Fig. 73. *Grinnellia americana.* x 1

Plants gregarious; individual plant consisting of a simple, erect blade, 1-5 dm long or longer, 4-10 cm wide, with a strong midrib, and a short stalk. Margins entire, though sometimes proliferous from the base. Blades a translucent pink color.

Collected from rocks in shallow water along outer beaches of St. Petersburg Beach.

Hypoglossum tenuifolium (Harvey) J. Agardh
Taylor (1960), p. 545.

Plants to 2 cm tall; habit spreading; blade proliferously branched from the midrib, main blades 3-5 mm wide, flat and oblong, midrib inconspicuous. Segments tapering, but tips obtuse. Colorless, light pink, or green in color.

Growing on mangrove roots at midlittoral level.

Taenioma nanum (Kützing) Papenfuss
As *Taenioma macrourum* Thuret, Taylor (1960), p. 548; Papenfuss (1951), p. 179.

Plants to 1 mm; with a creeping, polysiphonous main axis of four pericentral cells; attached by uncorticated rhizoids. Flat, determinate branchlets arise on the upper side of the prostrate axis, these blades are 60-75 μm wide and 240-400 μm long. Indeterminate branches alternate with determinate blades on every fourth to sixth segment, the determinate blades having four pericentral cells.

Common in tidal pools at Point of Rocks (Eiseman, in preparation).

Family Rhodomelaceae

Acanthophora muscoides (Linnaeus) Bory
Taylor (1960), p. 619.

Plants bushy, 6-16 cm tall. Irregularly branched, both the main axes (2.5 mm) and the spur branches beset with short (1 mm), spirally arranged spines. In cross section appearing parenchymatous. Found occasionally as dense clumps attached to rocks in heavy surf.

Acanthophora spicifera (Vahl) Børgesen
Fig. 74

Taylor (1960), p. 620.

Plants sparsely branched, to 25 cm tall; branches heavily clothed in upper regions with short spurlike branches. Spines present only on spur branches, the main axes lacking spines. Plants pale yellow brown to purple in color.

Common in shallow water, frequently straw colored and forming loose mats.

Fig. 74. *Acanthophora spicifera.* A, *x 7.5* B, *x 15* C, *x 0.75*

Bostrychia binderi Harvey
   Fig. 75
   Taylor (1960), p. 598.
   Plants purple to black, prostrate or erect, to 2 cm tall; the branches elongate and pinnate, the ultimate branchlets monosiphonous at the tips but polysiphonous at the base; main axis, including prostrate portions, polysiphonous and corticated; attached by a massive rhizoidal holdfast developed from a specialized branch.
   On mangroves, on buoys and rocks in the upper littoral level.

Bostrychia calliptera Montagne
   Joly (1954), p. 63; Taylor (1960), p. 597.
   Plants arising from a specialized holdfast as with *B. binderi,* except that the holdfast is found on the lower side of a major branching system; cortication rhizoidal and not parenchymatous; branchlets arranged in a pinnate pattern and polysiphonous to the apical cell; feathery in structure.
   Collected on mangrove roots in Tampa Bay (Rehm, in preparation).

Bostrychia moritziana (Sonder) J. Agardh

Taylor (1960), p. 596.

Plants ecorticate and with almost completely uniseriate terminal branchlets. Branching lax but basically pinnate; primary rhizoidal clusters replacing a branch. Plants reaching 3-4 cm.

Common on prop roots of *Rhizophora mangle* at the mouth of the Weeki Wachee River.

Bostrychia radicans Montagne

Including *B. rivularis* Harvey, Post (1936); Joly (1954), p. 58.

Plants compact to loosely branched, mosslike; reaching 2-3 cm in height; lower branches spreading, somewhat dichotomous; upper branches pinnate to bipinnate and polysiphonous; the last few segments of ultimate branchlets monosiphonous. Holdfasts developed from transformed branches and replacing a branch. Main axis of six to eight pericentral cells, branches four pericentral cells.

Common in upper littoral zone, especially on mangrove roots.

Bostrychia scorpioides (Gmelin) Montagne var. **montagnei** (Harvey) Post

Joly (1954), p. 62; Post (1936), p. 9; as *B. montagnei* Harvey, Taylor (1960), p. 598; Dawes (1967), p. 71.

Plants to 7.5 cm tall; branching pinnate, with no distinction between secondary and primary axes; main axes 650 μm in diameter, five pericentral

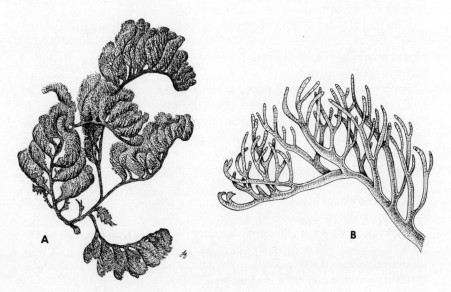

A   B

Fig. 75. *Bostrychia binderi.* A, x 2   B, x 11

cells, and five to seven layers of corticating cells; branchlets 70 μm in diameter, ecorticate and polysiphonous; terminal segments monosiphonous.

Growing on prop roots of red mangrove in sheltered areas.

Bostrychia tenella (Vahl) J. Agardh
Taylor (1960), p. 599.

Plants forming dense clusters; pinnately branched to several orders, older portions may be loosely branched; main axes 6 to 8 pericentral cells and corticated with three to four layers of parenchymatous cells; terminal branchlets monosiphonous, 25 to 35 cells long, cells about 20 μm in diameter.

Common on rocks, jetties, and mangrove pneumatophores.

Brongniartella mucronata (Harvey) Schmitz
Taylor (1960), p. 590.

Plants with a basal disk and a main axis with five pericentral cells; the axis corticated. Naked below and coated with short monosiphonous ramuli above.

Dredged off Egmont Key near buoy No. 2, in 25 m of water.

Bryocladia cuspidata (J. Agardh) De Toni
Taylor (1960), p. 586.

Plants to 8 cm tall, black; axes of eight pericentral cells, ecorticate; branches up to 1 mm long, simple, recurved-spreading and stiff; branches tapering from base to apex and closely investing the main axes.

Collected at Point of Rocks on limestone (Eiseman, in preparation).

Chondria cnicophylla (Melvill) De Toni
Taylor (1960), p. 614.

Plants loose to entangled, reaching 20 cm in length; attached by a discoid holdfast; main axes 0.7-1.2 mm in diameter, branching at right angles and alternate; branchlets 2-5 mm long, sharply contracted at the base, tips blunt; growing areas sunken, trichoblast evident in young branchlets; tetrasporangia scattered throughout fertile branchlets.

Attached to stones and shell in shallow-water grass flats.

Chondria collinsiana Howe
Taylor (1960), p. 617.

Plants erect, to 8 cm tall, branching alternate and spreading, main axis to 0.8 mm in diameter; cortication not obscuring end-wall thickenings of pericentral cells. Branchlets obtuse with tufts of brownish yellow trichoblasts at the apex; branchlets contracted at the base.

Plants collected during winter months on sea grasses near Anclote Key (Ballantine, 1972).

Chondria dasyphylla (Woodward) C. Agardh
Taylor (1960), p. 616.

Plants bushy, 10-20 cm tall, dark purple to light brown in color; main axes 1.0-1.5 mm in diameter; branchlets contracted at the base and club shaped; branchlet tips obtuse with growing areas sunken; each branchlet with a conspicuous tuft of trichoblasts; tetrasporangia formed in distal portions of fertile branchlets.

Growing on shell and stone in 1-5 m of water, usually a deep red color.

Chondria floridana (Collins) Howe
Taylor (1960), p. 616.

Plants up to 15 cm tall, more commonly 7-10 cm in length; pink to red in color; branching regular and alternate, giving a pyramidal aspect to plant. Main axis 2-3 mm in diameter, polysiphonous structure visible in cross section. Branchlets blunt with sunken apices and only occasional tufts of trichoblasts; branchlets not constricted although may be slightly tapered at base.

The species, with its characteristic features of nonconstriction of the branchlets and alternate branching, was found once, in May 1973, in great abundance in about 12 m of water due west of Anclote Key.

Chondria leptacremon (Melvill) De Toni
Taylor (1960), p. 615.

Plants erect, to 15 cm tall, slender in appearance, with more than one main axis; axes sparingly divided, main axes 0.22-0.32 mm in diameter; branchlets numerous, club shaped, tapering from blunt apex to the base. Apical cell immersed.

Found in 2-5 m of water in grass beds.

Chondria littoralis Harvey
Taylor (1960), p. 612.

Plants 10-20 cm tall, main axes 1-2 mm in diameter; older plants losing branchlets below; ultimate branchlets spindle shaped and contracted at base, 0.5 mm in diameter, 1-15 mm long; apical cell may be visible through a tuft of trichoblasts; tetrasporangia distal in fertile branchlets.

Common on rocks and shells in 2-5 m of water in grass beds.

Chondria sedifolia Harvey
Taylor (1960), p. 615.

Plants bushy, to 10 cm tall, usually a faded straw brown color, and firm in texture; main axes giving rise to several large, spreading branches, to 2.0 mm in diameter; ultimate branchlets sharply contracted at the base, club shaped; tip of branchlet truncate and with a sunken apical growing point; trichoblasts prominent at apex; tetrasporangia near ends of fertile branches.

On rocks and shells in grass flats.

Chondria tenuissima (Goodenough *et* Woodward) C. Agardh
    Fig. 76
    Taylor (1960), p. 613.

Plants 10-20 cm tall, main axes tough, simple, or with a few main branches; main axes 1.0-2.5 mm in diameter; ultimate branchlets spindle shaped, 2-7 mm long, and 0.2-2.5 mm in diameter; apex pointed and with an evident tuft of trichoblasts; tetrasporangia scattered in fertile branchlets.

Common in winter and spring on shells and stones among sea grasses, in 2-5 m of water.

Digenia simplex (Wulfen) C. Agardh
    Fig. 77
    Taylor (1960), p. 589.

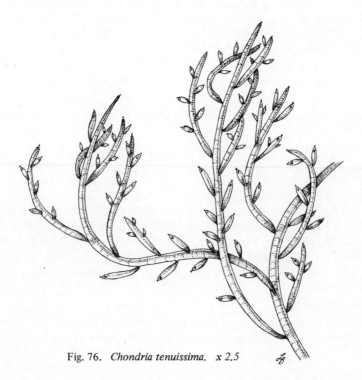

Fig. 76. *Chondria tenuissima.* x 2.5

Fig. 77. *Digenia simplex.* x 2

Plants 3-25 cm tall, wiry or cartilaginous in texture. Branching dichotomous to irregular, densely covered with slender, stiffened branchlets (3-5 mm long); branches may be denuded below. Brown to dull red in color.

Common on shells or stones in shallow water, often highly epiphytized by algae.

Falkenbergia hillebrandii (Bornet) Falkenberg

Taylor (1960), p. 571.

Plants small and filamentous, attached by branched, creeping rhizoids; erect portions usually alternately branched; main axis 30-70 μm in diameter; segments consisting of three pericentral cells, 1 to 2 diameters long. Tetrasporangia found in February.

Epiphytic on coarse algae (*Sargassum*), offshore Anclote Key in 9 m of water. Supposedly the tetrasporic phase of *Asparagopsis* (Nemalionales), which has not been collected along the west coast.

Herposiphonia secunda (C. Agardh) Ambronn

Taylor (1960), p. 604.

Plants to 2 cm, consisting of a creeping rhizome and erect axes, usually unbranched. Rhizome attached by rhizoidal cells; branching alternate, branchlet apices strongly curved toward the dorsal surface; older axes nearly straight, 75-150 μm in diameter; main axes with seven to eight pericentral cells; rhizome showing a regular pattern of branching, with an erect branchlet every five to eight segments.

On coarse algae, especially *Caulerpa,* in depths from 2-5 m. Recent work (Eiseman, in preparation) suggests combining this species with *H. tenella.*

Herposiphonia tenella (C. Agardh) Ambronn
Taylor (1960), p. 604.

Plants small and closely attached to substrate; branching of main axes alternate; prostrate; rhizomes producing straight, erect branches; axes 100-150 µm in diameter, every segment of rhizome having a branch or branch rudiment; main axes having 12 to 14 pericentral cells; branchlets with 35-45 segments, 65-90 µm in diameter.

On various algae, rocks, and mangrove roots.

Laurencia gemmifera Harvey
Taylor (1960), p. 624.

Plants to 15 cm tall, tough, cartilaginous, alternately branched throughout with branching tending to be in one plane (distichous). Branchlets short, 2 mm long, and not swollen at tip. Surface cells elongate, 40-130 µm long, quadrate in surface view with a small projection in the center or the entire cell pointed.

A characteristic plant in 15 to 30 m of water.

Laurencia intricata Lamouroux
Taylor (1960), p. 626.

Plants to 10 cm tall, matted, rose pink in color; branching alternate to irregular, main axes not visible, except in youngest portions; main branches 0.5-0.75 mm in diameter; branchlets cylindrical to club shaped to about 2 mm long.

Resembles *L. obtusa* but is smaller, found as drift and attached in quiet grass flats.

Laurencia microcladia Kützing
Taylor (1960), p. 627.

Plants to 10 cm tall; wiry, branching alternate, with pyramidal form; main axes 0.25-1.50 mm in diameter, ultimate branches 0.2-0.45 mm; inner cell walls thickened and visible through the thallus.

Uncommon, attached to rocks in the littoral zone.

Laurencia obtusa (Hudson) Lamouroux
Taylor (1960), p. 626.

Plants to 20 cm tall, usually matted; main axes sparingly branched, alternate, 0.75-1.50 mm in diameter; all branches bearing short, truncate branch-

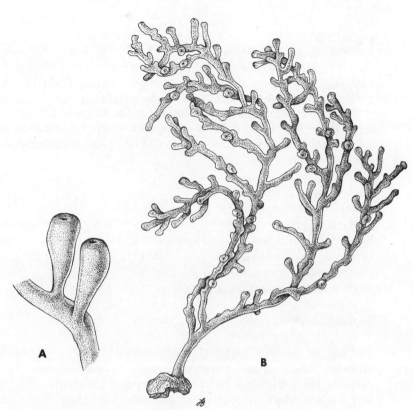

Fig. 78. *Laurencia poitei*. A, *x 5* B, *x 1.25*

lets with sunken apices; branchlets are 0.5-0.75 mm in diameter and only slightly enlarged at the tips.

A common red alga in offshore grass beds. Species distinctions are difficult to see between this and *Laurencia intricata*.

Laurencia papillosa (Forsskål) Greville
Taylor (1960), p. 623.

Plants densely clustered, to 5 cm tall, cartilaginous; main axes alternately branched, 1-2 mm in diameter; axes covered by short, truncate to tuberculate branchlets; in cross section cortical cells radially elongate.

Occasionally collected on exposed rocks.

Laurencia poitei (Lamouroux) Howe
Fig. 78
Taylor (1960), p. 625.

Plants to 10 cm tall, somewhat pinkish; the main axes alternately branched, branches cartilaginous, 1.0-1.5 mm in diameter; ultimate branches club shaped with truncate tips; branchlets are usually attached to flattened portions of the axes; cortical cells cuboidal.

A common alga of sea grass beds and rocks. Often highly epiphytized; forms dense balls up to a meter in diameter which roll along the bottom. Another form has elongated branchlets (4-5 mm long) and is found in deeper water (20-40 m).

Lophosiphonia cristata Falkenberg
Taylor (1960), p. 606.

Plants of prostrate filaments, often entangled; apices of rhizome and erect branches curved downward; main axes 90-110 μm in diameter; erect branches on dorsal side; trichoblasts forming a crest on the curved, erect filaments; pericentral cells from 6 to 12.

On *Sargassum* in 9 m off Anclote Key.

Lophosiphonia saccorhiza Collins *et* Hervey
Taylor (1960), p. 606.

Plants of prostrate filaments, 50-70 μm in diameter; segments 1 to 2 diameters long; rhizomes attached by large, elongate rhizoids, 160 μm in diameter and 600 μm long; axes having four pericentral cells; erect filaments to 2 mm tall, 25-45 μm in diameter, trichoblasts abundant, deciduous. Erect filaments sharply contracted at the base.

Epiphytic on *Codium*.

Murrayella periclados (C. Agardh) Schmitz
Taylor (1960), p. 593.

Plants purple, with a creeping rhizome and erect branches, having four pericentral cells. Branchlets radial in arrangement, monosiphonous except at the base (where polysiphonous).

Common at Siesta Key in crevices of coquina stone of the littoral zone.

Polysiphonia denudata (Dillwyn) Kützing
Taylor (1960), p. 580.

Plants 10-25 cm in length; deep red in color; main axes consisting of six (rarely five or eight) pericentral cells; segments of main axes 400-800 μm in diameter, 250-500 μm long, ecorticate; upper branches having four to six pericentral cells with segments 170 μm in diameter and 400-900 μm long; trichoblasts not obscuring a prominent apical cell.

Common on rocks and algae, in brackish water as well as the open coast.

Polysiphonia echinata Harvey
  Taylor (1960), p. 579.
  Main axis 6-10 cm tall, somewhat rigid and alternately branched, four pericentral cells; ecorticate or lightly corticated; coarse, main axes with closely spaced, spinelike determinate branchlets, about 1 mm long.
  Epiphytic on sea grasses at Anclote River, in brackish water (Ballantine, 1972).

Polysiphonia ferulacea Suhr
  Taylor (1960), p. 578.
  Plants to 15 cm tall, bushy, dichotomously branched above, the branches somewhat clustered; main axes of four pericentral cells, uncorticated; segments 200-300 μm in diameter, shorter than broad.
  Uncommon, on rocks in lower littoral level.

Polysiphonia gorgoniae Harvey
  Taylor (1960), p. 576.
  A small plant (to 3 cm), with a discoid base; main axes sparingly and dichotomously branched; segments consisting of four pericentral cells, ecorticated; trichoblasts obscuring the apical cell.
  Uncommon, epiphytic on sea grasses.

Polysiphonia hapalacantha Harvey
  Taylor (1960), p. 579.
  Plants densely branched, tufted, filaments to 15 cm tall; branching dichotomous below, less regular above; pericentral cells four, lightly corticated on main axes; main axes beset with slender, simple, spinelike branches, organized in four directions from the main axes.
  Common on rocks in littoral zone, appearing as a black, entangled mosslike plant.

Polysiphonia havanensis Montagne
  Taylor (1960), p. 577.
  Plants to 9 cm, alternately branched, apices with numerous trichoblasts; main axes 50-90 μm in diameter; segments with four pericentral cells, ecorticate, 1.5 diameters long in older portions; branches arising in axils of trichoblasts.
  On rocks and shells in shallow water, uncommon.

Polysiphonia howei Hollenberg
  Taylor (1960), p. 582.

Plants small and creeping; segments 100-170 μm in diameter, shorter than broad; pericentral cells 10 to 16; erect branches 2-5 mm tall, 70-150 μm in diameter; trichoblasts abundant near tip, leaving scar cells.

Collected on rocks, shells, and sea grasses, in shallow water.

Polysiphonia macrocarpa Harvey

Taylor (1960), p. 578.

Plants densely tufted, about 2 cm tall, rose red in color; main filaments 100-150 μm in diameter, 150 μm long; four pericentral cells and ecorticate; trichoblasts present; branching alternate.

Growing on pneumatophores in the lower littoral zone.

Polysiphonia opaca (C. Agardh) Moris *et* De Notaris

Taylor (1960), p. 583.

Plants forming large black cushions and drying to a black purple. Branching alternate mostly; widely divergent branchlets short, 1-5 mm long; trichoblasts common; segments with 14 to 20 pericentral cells; pericarps sessile and globular.

In tide pools at Point of Rocks (Eiseman, in preparation).

Polysiphonia ramentacea Harvey

Taylor (1960), p. 580.

Plants to 10 cm tall, tufted, a deep purple color; main axis appearing coarse; pericentral cells four, corticated in older portions; main axes covered with short (2-5 mm), determinate branchlets, which become shorter at the apices.

On other algae and rocks.

Polysiphonia subtilissima Montagne

Fig. 79

Taylor (1960), p. 575.

Plants to 15 cm tall, dark purple to straw colored; arising from a creeping base, branching mostly dichotomous; segments of main axes 90 μm in diameter and about 130-200 μm long; with four pericentral cells, ecorticate; apical cell conspicuous, with or without persistent trichoblasts.

Plants forming dense colonies on rocks or as loosely matted clusters on bottom of lagoons and tidal streams, as well as areas of higher salinities.

Polysiphonia tepida Hollenberg

Taylor (1960), p. 581.

Plants forming a dark reddish mat; filaments to 120 μm in diameter, erect filaments to 8 cm; ecorticate, with seven to eight pericentral cells; branches

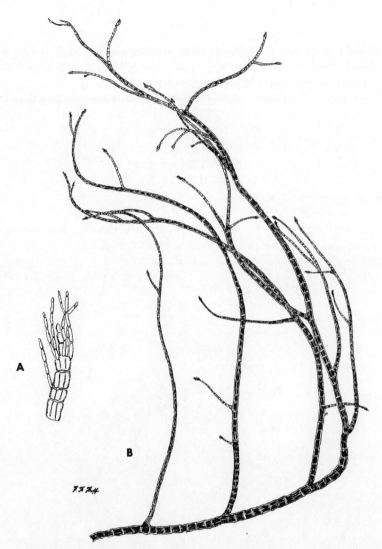

Fig. 79. *Polysiphonia subtilissima*. An enlargement on the left shows the apical cell with surrounding trichoblasts. A, x 10 B, x 0.5

arising at a basal fork of a trichoblast; branching alternate and distichous
Growing on stones in the sublittoral fringe; a common plant.

Wrightiella tumanowiczi (Gatty) Schmitz
Taylor (1960), p. 592.
Main axis attached by a discoid base; plants bushy, and with no primary axis; subdichotomous to alternately branched; to 75 cm long; having four

pericentral cells and heavy cortication. Branchlets pigmented and monosiphonous, forming dense coverings at the tips of branches. Monosiphonous branchlets usually found attached to spur branches.

Apparently a common plant offshore during winter, collected as free-floating plants in 9 m off Anclote Key.

Family Dasyaceae

Dasya corymbifera J. Agardh
    Taylor (1960), p. 559.
    Plants to 10 cm tall, branching alternate and pyramidal; main axes losing filaments below and becoming denuded, axes to about 0.75 mm in diameter; monosiphonous ramuli spirally organized.
    Found attached to shells offshore Tampa Bay channel in 15 m of water.

Fig. 80. *Dasya pedicellata.* x 1

Dasya pedicellata (C. Agardh) C. Agardh
Fig. 80
Taylor (1960), p. 562.

Plants light to deep red in color, erect, 2-7 dm tall, sparingly, alternately branched; branches 2-6 mm in diameter, often denuded. Main branches densely covered with slender branchlets (ramuli), 4-7 mm long. Branchlets randomly dispersed.

A common plant in winter-spring. A recent name change by Dixon and Irvine (1970) indicates this species should be called *Dasya baillouviana* (Gmel.) Montagne.

Dasyopsis antillarum Howe
Taylor (1960), p. 564.

Plants large, coarse, dark red in color, 4-10 cm tall. Irregularly branched below to alternate or subdichotomous above. Main branches slightly compressed, 1.0-2.5 mm in diameter; branches 1-3 mm long and covered with dichotomous ramuli.

Uncommon, apparently a plant of deep water, found once in offshore dredging.

8

Marine plant communities of the west coast

The west coast of Florida supports a rich and varied marine flora that consists of both angiosperms and algae. Three distinct marine angiosperm communities can be found along the west coast: sea grass beds, mangrove swamps, and salt marshes. In addition to angiosperm communities, lithophytic algal communities are found on rock jetties, pilings, and outcroppings of limestone, both sublittoral and littoral. This chapter describes these four plant communities, summarizing the most common angiosperms and zonation of both algae and angiosperms. A description of the general habitat and the location of a specific site for each of the four communities are also included. Physical data (Table 2) as well as ecological data (Table 3) are summarized for the four sites.

Table 2. **Physical Data**

Site	Salinity	Temperature	Substrate	Transparency	Wave Action (Average Wave Height)
Point of Rocks	32 to 38 0/00 Stable	20 to 30°C Seasonal	Shell hash, coastal limestone	Variable, depends on wave action	Moderate, 0.5 to 1 m in height
South end of Skyway	15 to 30 0/00 Estuarine, variable	16 to 32°C Seasonal, variable	Mud and sand, fibrous root system	1 to 2 m, usually turbid	Low, less than 0.25 m
Southside of Mullet Key	28 to 35 0/00 Semistable	18 to 30°C Seasonal	Sand, held by grass rhizomes	2 to 3 m, variable, depends on wave action	Low to moderate, less than 0.5 m
Mouth of Weeki Wachee River	5 to 20 0/00 Estuarine, variable	5 to 35°C Seasonal, variable	Mud, held by marsh plant roots	Turbid, less than 1 m	None, protected

Table 3. **Community Zonation**

Community	Supralittoral zone (spray zone, above high tide)	Littoral zone (region of tidal fluctuation)	Sublittoral zone (below lowest tide)
Point of Rocks	Rocks black to brown, *Calothrix* *Schizothrix* *Microcoleus* *Oscillatoria*	Tufted algae, *Cladophora* *Centroceras* *Bostrychia* *Bryopsis* *Hypnea*	*Caulerpa* *Acanthophora* *Ceramium* *Padina* *Sargassum* *Pterocladia*
South end of Skyway	Mangroves: *Laguncularia* *Conocarpus,* dead leaves, grass *Salicornia* if open enough *Entophysalis* *Microcoleus* on root bases, shells	Mangroves: *Avicennia* *Rhizophora* Filamentous algae on mud: *Cladophora* *Boodleopsis* *Vaucheria* Algae on roots: *Bryopsis* *Centroceras* *Caloglossa*	Limit of mangroves, oyster shells, prop roots: *Bostrychia* *Enteromorpha* *Rhizoclonium* *Caloglossa* *Polysiphonia*
Southside of Mullet Key	Sublittoral only	Sublittoral only	Sea grasses: *Thalassia* *Syringodium* *Diplanthera* Algal epiphytes: *Spyridia* *Fosliella* *Chondria*
Mouth of Weeki Wachee River	Marsh angiosperms: *Borrichia* *Distichlis* *Salicornia* Few blue-green algae	Marsh angio- sperms: *Juncus* *Spartina* Algae: *Batophora* *Chaetomorpha* *Centroceras* *Anacystis* *Oscillatoria*	No marsh angio- sperms; algae on bare mud, shells: *Batophora* *Polysiphonia* *Hypnea* *Anadyomene*

LITHOPHYTIC ALGAL COMMUNITIES

Point of Rocks (Fig. 1) is a shell hash or coquina rock outcropping found on the west side of Siesta Key, about five miles south of Sarasota. The outcropping extends for about 2 kilometers; it is about 100 meters wide, running from above high tide to about 3 meters below low tide. Remnants of a typical sand-strand community of sea oats (*Uniola*), sea grape (*Coccoloba*), Spanish bayonet (*Yucca*), and other plants can still be found growing above the high tide mark. The limestone is terraced by broken lines as it gradually slopes down to the flat sand bottom about 3 meters below low tide.

Species Description. The plant community exhibits zonation typical of a rocky exposed coast because of the heavy wave action (Table 3). The supralittoral zone consists of blue-green algae, especially *Microcoleus* and *Calothrix*, which form black to brownish regions on the rocks. The blue-green algal association is characteristic of the spray zone, and it is usually at or just above the barnacles. The blue-green algal association grades into a *Centroceras* association in the lower littoral region.

The *Centroceras* association of the lower littoral zone is characterized by filamentous turf algae (*Bryopsis, Ceramium, Chaetomorpha, Bostrychia*), as well as other larger morphologies (*Acanthophora* and *Hypnea*). The algae have a tufted to stunted appearance, possibly because of moderate to heavy wave action and scouring by sand. A comparison of plants (of the same species) taken from exposed and protected sites supports this observation.

The largest variety of algae are found in the lower littoral zone. A *Caulerpa* association is found consisting of three species: *Caulerpa prolifera, C. mexicana,* and *C. racemosa. Hypnea, Cladophoropsis, Gracilaria, Chaetomorpha, Spyridia, Pterocladia,* and *Cladophora* are also present in abundance, as are many other species. Sea anemones are also found in this zone.

The *Caulerpa* association extends into the sublittoral zone, but with a reduction in variety and quality of species because of water turbidity. At 2 meters, occasional plants of *Padina* and *Sargassum,* as well as *Ulva, Pterocladia,* and *Caulerpa* are found. Animals are conspicuous, including crustaceans, bryozoans, and *Cliona,* the bright orange to bright red boring sponge. Octopi are frequently encountered in rock crevices.

Habitat Description. The physical data presented in Table 2 indicate an open coastal environment with moderate wave action and stable salinity, temperature, and pH. The outcropping appears to be recent beach rock (less than five thousand years old) formed by cementation of quartz sand and marine invertebrate shells by calcite deposited through organic processes. Wave action is moderate (Tanner, 1960)—0-0.5 meter or rarely 1 meter in height.

Location. Point of Rocks at Siesta Key can be reached by driving south on Highway U.S. 41 from Sarasota for about five miles and then turning right on Florida State Road 72 (going west). After about one mile, State Road 72 terminates at county road 789. Turn left (going south on Siesta Key) on county road 789 and proceed for about one-half mile until a small sign on the right indicates the Point of Rocks road. Turn right, proceeding for one block, and park at the public access trail to the beach. The limestone outcropping begins on the south side of the beach.

MANGROVE SWAMPS

Mangrove swamps are frequently encountered since they are the major form of coastal vegetation along the west coast of Florida from Tampa Bay southward in the littoral zone. The forest-swamps may form large regions of scattered islands, such as the Ten Thousand Islands of south Florida, or a more distinct, narrow belt along the west coast. Most ecologists consider the mangrove swamp a temporary association that will be replaced ultimately by an upland, nonsaline plant community (Humm, 1973d).

One species, the black mangrove, occurs as stunted shrubs as far north as the Mississippi Delta. Davis (1940, 1943) is apparently correct in his opinion, however, that the mangroves are frost sensitive and so are mostly limited to coastal sites from Tampa Bay southward. A number of excellent mangrove islands (Fig. 2) can be reached by wading on the south end of the Skyway on Highway U.S. 19, at the mouth of Tampa Bay.

Species Description. Mangrove communities are dominated by three dicot trees: *Rhizophora mangle* (red mangrove), *Avicennia germinans* (black mangrove) and *Laguncularia racemosa* (white mangrove). A fourth dicot tree, *Conocarpus erecta* (buttonwood), is usually found as a component of the maritime vegetation and will not be considered here (but see Humm, 1973d). The term "mangrove," as mentioned by Davis (1940), is ecological rather than taxonomic. The principal adaptations in common with a littoral habitat are the ability to grow in waterlogged soil, vivipary of the fruit (germination on the tree), and xerophytic adaptations.

Probably the most characteristic feature of all mangroves are the anatomical adaptations for prevention of water loss (xerophytic). Such adaptations, characteristic of desert plants, demonstrate the presence of rigorous conditions in the littoral area. Studies of the anatomy of the mangrove leaves, the specialized morphology of the root systems, and the adaptations for salt removal all support the previous statement.

The leaf anatomy of the mangroves is especially indicative of xerophytic adaptations. Generally, the epidermal layer has a thickened cuticle, and stomata are restricted to the lower side of the leaf. A hypodermis of two or three layers of tannin-containing cells may be found adjacent to the epidermis on both sides of the leaf. Hydrocytes (water storage cells) and cystoliths (crystal-containing cells) are common in the mesophyll of the leaves. The mesophyll is differentiated into spongy and palisade mesophyll tissue.

The common mangroves are:

1. *Rhizophora mangle* Linnaeus (Red mangrove). An evergreen shrub or tree (to 20 meters) with aerial roots arising from trunk (prop roots) and branches (drop roots). Leaves 5-15 cm long, opposite, leathery, entire, and persistent. Apical buds with a distinct pointed sheath. Fruits, long and green, germinating on tree (viviparous).

The size of the plant is related to the frequency and severeness of frosts, with only shrubby plants found north of Tampa Bay (Davis, 1943). Plants form an almost impenetrable barrier in lower littoral levels along the bay and river shores in salinities of 5 to 20 0/00. The prop roots offer a unique substrate to littoral algae.

2. *Avicennia germinans* (Linnaeus) Stearn (Black mangrove). Trees or shrubs (to 25 meters), bark dark and scaly; opposite leathery leaves of light to dark green, 3-10 cm long, elliptic in shape. Distinct erect roots (pneumatophores), which project from the substrate to 20 cm, are characteristic of the species. Leaves form salt crystals at sites of hydathodes on their upper surface. Fruits viviparous.

A common plant found in the mid-littoral zone, above *Rhizophora*. The root system of *Avicennia* is complex, usually consisting of five types of roots: (1) primary adventitious roots that grow down from the base of the main stem; (2) horizontal roots that radiate out from the stem from which (3) anchoring roots direct downward into the substrate and (4) pneumatophores project upward from the substrate. Absorbing roots (5) extend horizontally from the pneumatophores, forming a fine, fibrous root system. An aerenchyma tissue forms the cortex of the entire root system, probably permitting gaseous exchange. The pneumatophores tend to carry distinctive algal epiphytes, such as *Bostrychia* and *Caloglossa*.

3. *Laguncularia racemosa* Gaertner (White mangrove). Shrubs or trees (to 20 meters) with opposite evergreen leaves. The leaves are 4-7 cm long, oblong, simple and leathery, with cordate tip (indented at apex of leaf), and a pair of glandular nodes at base of the blade.

Plants are found in the upper littoral or supralittoral zone, typically above red and black mangroves. This species appears to mark the upper level of salt water.

Habitat Description. The mangrove stands found along the Skyway are well developed, particularly the south sides of the islands. The north sides of these islands are buffeted by low to moderate wave action (Tanner, 1960), which uproots the mangroves. Physical data taken in the mangrove swamp (Table 2) indicate an estuarine environment with a wide range in salinity and temperature. The substrate, which consists of fine sand and organic mud, remains in suspension for long periods after slight disturbances, thus lowering the water transparency (0.5-1.0 meter).

The substrate in the supralittoral zone is rather barren, consisting of sand and shell with dry leaves scattered over it. The region is usually dominated by the buttonwood (*Conocarpus*) and white mangrove (*Laguncularia*) and occasionally some black mangrove plants (*Avicennia*). Blue-green algae can be found on the shells and soil (Fig. 3).

The pneumatophores of the black mangrove mark the littoral zone. The substrate of the littoral zone is black, composed mostly of silt and is high in organic content. The dominant mangrove is *Avicennia,* but there are occasional red mangroves. Filamentous algae (*Cladophora, Chaetomorpha, Boodleopsis, Vaucheria, Centroceras*) are found matted together on the mud. A number of other algae are usually found growing on the pneumatophores; they exhibit a vertical zonation. The uppermost portion of the pneumatophores have blue-green algae (corresponding to the supralittoral). The intermediate region has *Rhizoclonium* and *Enteromorpha* (upper littoral). *Bostrychia* and *Caloglossa* are found on the lower littoral to upper sublittoral regions of the pneumatophores.

The red mangrove becomes the dominant tree at the outer edge of the swamp (Fig. 81). In this outer region, oysters, bryozoans, and small sponges can be found attached to the prop roots of *Rhizophora*. The major algal

Fig. 81. The edge of a mangrove swamp showing the prop roots of *Rhizophora,* the red mangrove (left side), and the pneumatophores of *Avicennia,* the black mangrove (right side). Both algae and oysters are attached to the roots. Photograph taken at low tide.

epiphytes of the mangroves diminish in number. *Ulva, Monostroma, Centroceras,* and *Polysiphonia* appear as epiphytes on the prop roots and shells.

In summary, the mangrove swamp is characteristic of a subtropical to tropical estuarine environment. It is primarily littoral in occurrence and found in locations of low wave energy. The presence of aerial roots (prop and drop roots, pneumatophores) appears to be important in transpiration (Bowman, 1917) and in land formation (Davis, 1940; 1943). The prop roots of the red mangrove trap debris and develop substrate while the fibrous root systems of pneumatophores and the pneumatophores themselves further develop a peatlike substrate (Davis, 1940).

The variety of algal species is low, but the forms present apparently also play a role with the mangroves in land formation (Dawes, 1965). The filamentous algae form mats on the dark substrate, trapping finer soil particles and plant debris. Consequently, the substrate level is increased so that the less salt-water tolerant species, such as *Laguncularia* and *Conocarpus,* may germinate.

Location. The mangrove stands described are present on the west side of Highway U.S. 19 on the causeway just south of the Skyway bridge.

⸱ SEA GRASS COMMUNITIES

Sea grass beds form extensive sublittoral plant associations along the west coast of Florida (Humm, 1956). Although not true grasses (Humm, 1973c), these monocots resemble grasses in their general morphology, having long leaves, rhizomes, and fibrous root systems. Most experts consider sea grasses to have evolved from freshwater monocots that spread from the mouths of rivers into the ocean (Arber, 1920). Recently, den Hartog (1970) has argued that these hydrophytic monocots are evolved from xerophytic salt marsh plants.

Regardless of their evolution, the sea grasses have a very important environmental role in the sublittoral community (Humm, 1973c).

1. They serve as a sediment trap and stabilize the bottom sediments.

2. They carry on basic productivity that may exceed that of the benthic algae. Jones (1968) has determined that well-developed beds of turtle grass fix about 900 grams of carbon per square meter per year.

3. They serve as direct food sources for some herbivores (sea urchins, certain fish, sea turtles, manatees) and as an indirect food source in the form of detritus for other invertebrates.

4. They provide a habitat and refuge for numerous organisms including scallops, shrimp, crabs, and many fish.

5. They serve as a substrate for many invertebrate species and for algae (Humm, 1964; Ballantine, 1972).

Although somewhat patchy in distribution, the sea grass beds along the west coast of Florida probably occur in a band that may reach ten to fifteen miles in width. With regard to abundance, Phillips (1960a) has estimated that the common turtle grass (*Thalassia*) forms a bottom cover with a standing crop in Boca Ciega Bay of about 3,000 pounds per acre in dry weight of leaves.

A "typical" sea grass community would be found offshore, for example, at buoy No. 2 on the Homosassa River channel. A sea grass community off the south side of Mullet Key (Fort DeSoto Park, Pinellas County) is described here because it is reasonably accessible.

Species Description. The sea grasses are hydrophytes, that is, plants that grow underwater, and hence show great reduction in tissue complexity. Cross sections of sea grass leaves, for example, will demonstrate a very poorly developed vascular system (Thomlinson and Vargo, 1966). Furthermore, there is a substitution of aerenchyma for the typical palisade and spongy mesophyll tissues in leaves of land plants. An apparent loss of cuticle and stomata from the epidermal layer can be seen. Reproduction (pollination and flower formation) in these hydrophytes is also highly specialized (Rydberg, 1909; Bowman, 1923).

Six species, which are placed in five genera, can be recognized along the west coast of Florida:

1. *Thalassia testudinum* Koenig *et* Sims (Turtle grass). Leaves 2-5 in a cluster, linear and strap shaped, 5-30 cm long, 4-9 mm wide, rounded at apex. Stem a persistent rhizome, with leaf clusters arising at nodes.

It is the most common sea grass, and it forms dense beds in salinities of 25 0/00 to 40 0/00. It occurs from mean low tide to 30 meters in depth. Moore (1963) studied the distribution of this species, and Jones (1968) studied its primary productivity.

2. *Halophila baillonis* Ascherson (no common name). Plants small, consisting of two or three blades, sessile on rhizome. Blades obvate, three veined. Plants reaching 1-3 cm in height.

Common in depths of 10 to 60 meters, offshore from Tampa Bay, although occurring in shallow water in the Florida Keys.

3. *Halophila englemannii* Ascherson (no common name). Plants a series of erect shoots to 4 cm long, arising from a creeping, scaly rhizome. Leaves three-ribbed, 1-3 cm long, fleshy and arising at apex of erect stems in pairs or whorls of three to five.

Less abundant but frequent in salinities of around 20 0/00 to 40 0/00, usually in depths less than 5 meters. This species will grow mixed with *Thalassia* and *Syringodium*.

4. *Diplanthera wrightii* (Ascherson) Ascherson (shoalgrass). Plants growing in sand to mud bottoms with short, erect stems arising from rhizome. Leaves clustered on short stems; leaves narrowly linear, 3-40 cm long, 0.5-1.0 cm in width.

Common in salinities of 12 0/00 to 20 0/00, and may occur in higher salinities (to 40 0/00). Usually in depths of 0.2 to 10 meters. Phillips (1967) studied this plant, designating it by the name, *Halodule wrightii* Ascherson. Characteristic of sandbars in mouths of rivers, hence its common name, shoalgrass.

5. *Ruppia maritima* Linnaeus (Widgeon or ditch grass). Plant having small, flexuous leaves, 10-12 cm long and 0.2 mm to 1 mm wide; leaves arising from a simple to much branched rhizome that is poorly attached to substrate. The rhizome shows a characteristic alternate pattern of leaves at the growing tip.

Plants basically freshwater but common in salinities of 5 0/00 to 10 0/00; growing on mud and sand in the littoral and upper sublittoral zones.

6. *Syringodium filiforme* Kützing (Manatee grass). Plants with horizontal rootstock and terete leaves. Leaves to 35 cm long, about 0.5 cm in diameter, and with stipular (pointed) sheaths and bases.

Typically found in salinities of 20 0/00 to 40 0/00 and in 2 to 15 meters of water, mixed with *Thalassia*. A common offshore plant, often termed *Cymodocea manatorum* Ascherson.

Habitat Description. The distribution of sea grasses is influenced by depth, turbidity (Strawn, 1961; Phillips, 1962), and salinity (Phillips, 1962). In a study of sea grass distribution in Tampa Bay, Phillips (1962) indicated that *Ruppia* is a plant characteristic of low salinities (5 0/00 to 15 0/00) and shallow water. *Diplanthera* was found intermixed with *Ruppia* in the sublittoral regions where the salinities were 10 0/00 to 20 0/00. In salinities of 15 0/00 to 25 0/00, *Syringodium* became the dominant sea grass. *Thalassia* is dominant in salinities higher than 20 0/00 (Phillips, 1962), and it will form dense beds at depths exceeding 20 meters in the clear offshore waters or to about 3 meters in the turbid bay waters. The clear water and firm (limestone) substrate of offshore water offer an excellent environment for luxuriant growth of *Thalassia*, *Syringodium*, and *Halophila*.

Sea grass beds, as found on the south side of Mullet Key, range in depth from 0.5 to 10 meters, depending on water transparency. The physical data for the site selected (Table 2) indicate an estuarine environment, because it is near the mouth of Tampa Bay. The south side of Mullet Key, however, is open to southerly winds, causing a low variation in salinity and temperature. The substrate is hard sand. The grasses (Table 3) tend to form pure or mixed stands of *Thalassia*, *Diplanthera*, and *Syringodium*. Some of the more common algal epiphytes on the sea grasses include *Champia*, *Lomentaria*, *Polysiphonia*, *Acrochaetium*, *Fosliella*, *Hypnea*, *Spyridia*, *Cladosiphon*, *Ectocarpus*, and *Cladophora*. Some or all of these may be present at one time, but the brown algal epiphytes are typically found in the later winter months. Plants attached or free on the bottom include *Gracilaria, Hypnea, Ulva, Enteromorpha,* and *Anadyomene.* Offshore sea grass communities contain a richer algal flora, including siphonaceous green algae (*Acetabularia, Batophora, Penicillus, Udotea, Halimeda,* and *Caulerpa*), various red algae (*Laurencia, Eucheuma, Dasya, Chondria, Digenia*), and brown algae (*Sargassum, Padina, Dictyota, Pocockiella,* and *Sporochnus*). Sponges and soft corals are common in the offshore sea grass beds.

Location. After arriving on Mullet Key (Fort DeSoto Park), drive east on the Mullet Key road past the east swimming and picnic area. Park near the end of the road.

SALT MARSH COMMUNITIES

Salt marsh communities, like mangrove swamps, are plant communities of the littoral zone found in regions of low wave energy, such as embayed rivers and protected bays. Salt marshes are the dominant coastal vegetation north of Tampa Bay, where frosts are common; these marshes essentially replace the mangrove swamps.

Like the mangrove trees, the salt marsh plants exhibit xerophytic adaptations that are especially apparent in their leaf anatomy. Unlike the dicot trees forming the mangrove swamps, the dominant plants of a salt marsh are monocots. A representative marsh can be found at the mouth of the Weeki Wachee River (Fig. 3).

Species Description. Salt marsh plants are angiosperms, both dicots and monocots, with grasses and rushes being the most conspicuous. The first four plants in the following list are the most common dicots, and the last five plants are the most common monocots.

1. *Batis maritima* Linnaeus (Saltwort). A succulentlike shrub with opposite, pale green (glaborous) leaves on a prostrate stem. Strongly scented shrub. Flowers are formed in bracts, and the inflorescence forms a cone. The leaves are characteristic of the genus, being quite gray green and hairy.

The plant is commonly called a saltwort; it is typical of maritime providences in the upper littoral region of a salt marsh.

2. *Borrichia frutescens* (Linnaeus) DeCandolle (Sea daisy). Plants to 1 m tall with fleshy, opposite linear leaves borne on erect stems that become woody. Leaves to 7 cm long, may be finely toothed. The flower is a composite, and the fruit is woody when mature.

The species is very characteristic of the maritime zone, especially in the upper littoral regions of salt marshes.

3. *Salicornia virginica* Linnaeus (Glasswort). A fleshy shrub on which the

leaves are arranged in opposite pairs and are wrapped around the stem so that the green stem appears succulent and leafless. The plants have a procumbent to rhizomatous stem with short (5-20 cm), erect stalks.

Commonly called "glasswort" because of the crunchy sound when stepped on. The species occurs in the upper to mid-littoral zones of salt marshes and disturbed mangrove swamps. Also known as *S. perennis* Miller.

4. *Sesuvium portulacastrum* Linnaeus (Sea purslane). A succulent herb with a prostrate stem and opposite leaves. The leaves are spatulate and reddish green, 1-5 cm long. Small pink flowers arise from the axils of the leaves.

Plants are characteristic of the upper littoral zone of marshes, and a common sea-strand plant.

5. *Distichlis spicata* (Linnaeus) Green (Salt grass). A creeping grass with rigid, erect stems (15-20 cm tall) and overlapping leaf sheaths. The leaves have a distichous arrangement, hence the generic name.

A common salt grass, present along the entire west coast of Florida. It is very obvious as a member of the upper littoral region of a salt marsh.

6. *Fimbristylis castanea* (Michaux) Vahl (Sedge). A small (to 1 m tall), deep green, perennial sedge with involute leaves; blades narrow, rigid, 2-4 dm long. Spikelets are arranged in a rosette and become golden brown when in fruit.

A sedge characteristically found in the upper littoral zones of the salt marsh and along protected brackish sandy regions.

7. *Juncus roemerianus* Scheele (Black rush). Plants stiff, rigid, 1.5 m tall in large tufts, gray green to black in color. Leaves terete, channeled on adaxial side, elongate and septate.

The black rush is one of the most important species of the salt marsh community, based on numbers of individuals and amount of land occupied. From a distance a *Juncus* stand appears to be gray green in color, dominating the mid-littoral zone.

8. *Spartina alterniflora* Loisel (Smooth cordgrass). Plants with compressed to terete culms, to 2 m tall; leaves broad, flat (to 2 cm wide), arising from a spongy base. Leaves a bright green and in a distichous arrangement.

The plant occurs in the lower littoral to sublittoral zones of the salt marsh, forming a fringe.

9. *Spartina patens* (Aiton) Muhlenberg (Slender cordgrass). Culms 0.5 to 2 m tall; leaves narrow, to about 1 cm in width, relatively bright green in coloration. Rhizomes slender, slightly scaly.

The species is usually found mixed with *Juncus* or forming continuous, dense stands in the mid-littoral to upper littoral zone of salt marshes (Kurz and Wagner, 1954).

Habitat Description. Humm (1973*a*) has pointed out that one of the most characteristic features of salt marshes is the distinct zonation exhibited by the more common plants, which is controlled to a major extent by the periodic rise and fall of the tide. Jackson (1952) has also noted such factors as composition of the substrate, height above sea level, pH, and rate of percolation of salt water during high tide as well as amount of freshwater runoff.

Salt marshes are usually found along stable or emerging shorelines where accumulation of sediment in a shoalgrass community will permit a stand of *Spartina alterniflora* to develop. With the increase in sediments, *Spartina* may be replaced by *Juncus*, which may slowly give way to *Distichlis* if the soil is highly organic or to *Spartina patens* if a sandy soil is being deposited (Humm, 1973a). Humm (1973a) lists three major ecological functions in regard to the role of salt marshes; salt marshes are

1. The site of production of large quantities of organic matter (Heald, 1969).

2. The habitat of a number of species of algae, seed plants, invertebrate animals, and of a considerable number of birds.

3. A buffer zone between the marine and terrestrial vegetation.

The salt marsh found at the mouth of the Weeki Wachee River is gray green in color because of the rush *Juncus*. *Spartina alternifora* forms a band along the lower littoral region, especially along tidal channels (Fig. 3), while a number of halophytic plants, such as *Spartina patens, Distichlis, Borrichia, Sesuvium, Salicornia, Batis, Fimbrystylis,* and others, form the upper littoral and supralittoral transition zone (Table 3). The latter transition zone separates the littoral marsh and the Sabal Palm-Juniper hammocks. *Acrostichum,* a large fern with bright green leaves, is also present along the edges of tidal channels.

Algae are found as a mat on the mud (*Anacystis, Microcoleus, Oscillatoria, Cladophora, Chaetomorpha, Centroceras, Vaucheria*), on plant stems (*Bostrychia, Catenella; Polysiphonia*), or on shells and broken limestone (*Batophora Enteromorpha, Centroceras*). The filamentous algae form a mat over the substrate similar to that found in a mangrove swamp (Fig. 82). The lower littoral zone of tidal channels is essentially exposed mud, lacking both salt marsh plants and algae. In the tidal channels, *Ruppia,* along with such algae as *Batophora, Spyridia, Sargassum, Polysiphonia, Ulva,* and *Enteromorpha* is found. Freshwater algae, such as *Chara, Oedogonium,* and *Spirogyra,* can be found in the more brackish regions (1 0/00 to 5 0/00) of tidal channels.

The substrate of the salt marsh consists of mud held together by the fibrous root system (Table 2). The substrate has a pH of 6.5 to 7.2 because of the high organic content. Salinity is very low at the river mouth, ranging from less than 5 0/00 to 20 0/00. The higher salinity readings are obtained during the dry, winter months. Not only is there a wide range in salinity, but there is

Fig. 82. A view of the substrate of a salt marsh. The grasses present are *Spartina* and *Juncus*. The fibrous, matted appearance of this substrate seen at low tide is caused by a green algal mat consisting of *Cladophora* and *Boodleopsis,* as well as *Vaucheria.*

also a wide range in temperature. Such wide fluctuations in temperature and salinity limit the algal flora of a salt marsh. An excellent study of Florida salt marshes has been done by Kurz and Wagner (1954).

Location. Drive north from Saint Petersburg on Highway U.S. 19 to Weeki Wachee Springs and turn left, going west, on Florida State Road 50 to the mouth of the Weeki Wachee River. The drive down State Road 50 demonstrates the gradual change from highland pine communities through Sabal Palm-Juniper hammocks to the continuous salt marsh communities at the river mouth.

Glossary

abaxial: on the side away from the axis or branch

abscission: the natural separation of a portion of a plant

aculeate (aculeus): pointed or spinelike

adaxial: on the side toward an axis or branch

akinete: in the Chlorophyta and Cyanophyta, a thick-walled, vegetative cell produced in asexual reproduction

anastomose: united by cross connections, or closely adjoined at cell walls (see *Anadyomene*)

annular, annulate: ring shaped or arranged in rings

antheridium: sex organ producing flagellated male gametes (sperm)

anticlinal: perpendicular to the surface

apical cell: a distinct meristematic cell initiating growth at the apex of an axis or a branch

apiculate: with a point

appressed: pressed against main axis, having a small angle in branching

articulated: jointed, with segments

attenuate: narrow, gradually tapering to a point

basipetalous: with progressively younger portions from the base to the apex; usually with relation to spore production or sexual reproduction

bipinnate: twice pinnate

blade: an erect, leaflike, flattened portion of a plant

brackish water: a mixture of at least one-third freshwater and two-thirds or less seawater (salinity less than 10 0/00)

branchlets: the ultimate branches of a plant (ramuli, ramulus), cells containing plastids as opposed to colorless hairs (trichoblasts)

bulbous: swollen, ball shaped

calcified: stony, the cell walls usually impregnated with lime ($CaCO_3$)

carpogonium: female sex organ in the Rhodophyta, equivalent to an egg

carpospore: a spore resulting from proliferation of the carposporophyte

carposporophyte: a separate generation borne on the gametophyte in Rhodophyta; consists of haploid or diploid gonimoblast filaments and carposporangia

cartilaginous: hard and tough; plant having the feeling or texture of cartilage

cervicorn: branching more or less one-sided, suggesting the horns of deer

clavate: club shaped

coalescent: fused together; branchlets of *Struvea*

coccoid: single celled, spherical when isolated

coenocytic: not divided into cells: a filament without cross walls; essentially a continuous tube of cytoplasm

compressed: thallus laterally flattened

conceptacle: sunken "pits" containing reproductive bodies, such as tetrasporangia, cystocarps (of Corallinaceae), or antheridia and oogonia (of *Sargassum*)

constricted: having an abrupt narrowing of a filament or branch giving the appearance of an annular ring

cordate: heart shaped

cornute: crownlike, spine arrangement in *Hypnea cornuta*

corona inferior: a ring of shorter appendages arising from the stalk below the main ring of branches of the *Acetabularia* cap

corona superior: similar to the corona inferior but arising above the cap of *Acetabularia*

cortex: the tissue between the epidermis and medulla; if no epidermis is present, the region surrounding the medulla or the axial filament

cortication: covering cells on an axis or branch, from rhizoids or parenchyma

costate: a raised midrib forming a ridge, as in *Sargassum pteropleuron*

crenulate: the margin undulating

crosier: recurved, hook branch, as in *Hypnea musciformis* (*see also* hemate)

cruciate: type of division in which only two of four cells in a tetrasporangium are visible; planes of division perpendicular to one another

crustose: crustlike; in the case of algae, the term restricted to thin plants growing flattened against the substratum

cryptostomata: a cavity sunken in a thallus and containing hair cells only; especially common on foliar blades of *Sargassum*

crystolith: a specialized cell that contains a crystal, usually somewhat branched

cuboidal: cube shaped

cystocarp: a term used only in the Rhodophyceae to include the carposporo-phyte and any gametophytic tissue developed with it; usually the structure globose, protruding

deciduous: falling away; not persistent

decumbent: prostrate, but the apex bent upward

dentate: toothed

determinate: not growing beyond certain size, as in the side branchlets of *Crouania;* ultimate branches or branchlets may fall into this category if no further growth will occur

dichotomous: forked very regularly, the divisions of equal length

discoid: shaped like a disk, round and flat

distal: the tip or ultimate end, as in a branch tip

distichous: arranged in two vertical rows along the opposite sides of an axis or branch

distromatic: two cells in thickness (with reference to flattened blades)

dorsal: the upper surface of a dorsiventral (prostrate) plant

ecorticate: lacking outer covering (cortical) cells

endogenous: formed internally in plant thallus

endophyte: a plant growing within the tissues of another plant

entire: without divisions or lobes; sometimes refers to smooth margins

epidermis: the outermost cell layer of a plant

epiphyte: a plant growing on another plant

erose: margin (of blade) uneven, irregular; toothlike edge

fasciculate: a cluster of parallel filaments, stems, or branches

fastigiate: tapering to a point

fibrillar: made up of fibrils, strands woven together

filament: a branched or unbranched row of cells joined end to end; in the Cyanophyta the term filament includes the cell row (trichome) and sheath

filiform: filamentous or threadlike; edges of blade tapering to numerous fine strands (*Halymenia*)

flabellate: fan shaped

fleshy: soft and smooth in texture; somewhat cartilaginous; not slimy or gelatinous; firm

floats: small bulbous or spherical beads, as in *Sargassum*

foliaceous: bearing leaflike structures

foliar (foliar leaves): portions of an alga that resemble leaves of higher plants

forcipate: forked and incurved like a pair of forceps or pincers

fruiting: bearing sporangia, gametangia or cystocarps, or other reproductive structures

gametophyte: the sexual or gamete-bearing generation of a plant

gelatinous: like jelly or gelatin in texture

gland cell: a colorless cell with a birefringent appearance (see *Botryocladia*)

gonimoblast: a carposporangium-producing filament growing from the carpogonium or from an auxiliary cell of the Florideae

hair cell: a nonpigmented cell

haptera: a multicellular basal organ of attachment; usually cylindrical and much branched (see *Bostrychia*)

haustorium: a specialized filament penetrating living cells of a host plant; usually pertaining to parasitic plants

hemate: a hooked branch (*see also* crosier)

heterocyst: a dense, thick-walled cell, usually larger than adjacent vegetative cells as in *Lithothamnium* of the Rhodophyta and many members of the Cyanophyta; function unknown

heterothallic: male and female organs borne on separate plants

holdfast: attachment structure for an alga, one or more cells modified to anchor the plant

homothallic: male and female sex organs borne on the same plant

hyaline: colorless, transparent; lacking plastids or pigmentation

hydathode: structure for removal of salt water, a modified vein ending in leaves

hydrocytes: cells specialized to hold water, usually thin walled and large

hypothallus: the lower portion (tissue) of crustose or dorsiventrally flattened plants

indeterminate: the main growing axis of a plant; opposite of determinate and is expressed by unlimited growth

indusium: a membranous covering of a reproductive site

intercalary growth: meristematic activity occurring anywhere along a filament; except at the base of apex (as in the Corallinaceae)

internode: the region between two nodes; in a filament the part of a cell between the end walls

intertidal: occurring between the low and high tide marks

isodiametric: equalsided

lamina: leaves or blades of an alga

lanceolate: lance-shaped; broad in the middle and narrow at each end

lenticular: having the appearance of a lens

ligulate: strap shaped and short

linear: long and narrow, with sides parallel

lithophytic: a plant growing on rocky material

littoral: roughly the zone between highest and lowest tides, but the area can extend higher with wave action

mannose: a polymer of the carbohydrate mannan

matted: entangled or woven together

medulla: the central tissue of an internally differentiated thallus; can be filamentous or parenchymatous

membranous: sheetlike and flat, thin

meristematic: site of active cell division in a plant

micrometer (μm): a common metric unit of linear measurement used in microscopy, equals 1×10^{-6} or 0.000001 meter

midrib: the thickened longitudinal axis of a blade

mitochondria: cell organelles that carry respiratory enzymes and are site of Krebs cycle

moniliform: deeply constricted at regular intervals or at the nodes and hence resembling a string of beads

monopodial: plant with distinct main axis bearing lateral branches of smaller size

monosiphonous: having a central siphon (*see* uniaxial)

monosporangium: a sporangium containing one spore, thought to be a modified tetrasporangium of the Rhodophyta

monostromatic: consisting of a single layer of cells; with reference to membranous blade thalli (see *Monostroma*)

multiaxial: with a medulla of several parallel longitudinal filaments, each terminating in an apical cell

multiseriate: more than one cell in width, a filament two or more cells wide (*see also* pluriseriate)

muriculate: coarse or rough to feel due to presence of spines or irregularities

nemathecium: a wartlike structure containing reproductive organs; found on the surface of some Florideae

node: juncture between cells in a filament or between branches and main axes in multicellular plants

obovoid: pear shaped or shaped like a tear drop

oogamous: a plant in which gametic union takes place between an antherizoid and an egg

orbicular: spherical or rounded

organelle: membraned unit of the cell (plastid, mitochondria)

ostiole: the porelike opening in a conceptacle or pericarp

ovate: egg shaped, elliptical

ovoid: ovate

palmate: as in palm of hand, branching or lobing of equal importance

papilla: a short, nipplelike outgrowth

paraphyses: short, sterile filaments usually found adjacent to a reproductive organ

parasite: a plant borne upon or within another organism and deriving food from it

parenchyma: tissue composed of thin-walled living cells

parietal: adjacent to wall

pectinate: with lateral branches restricted to one side and set close together like the teeth of a comb

pedicel: stalk supporting a reproductive structure

pelagic: free-floating, not attached, planktonic

peltate: shield shaped, with stalk attached in center of blade

pericarp: the sterile envelope produced by the gametophyte around the carposporophyte in Florideae

pericentral cell: cells found surrounding the central filament which were ontogenetically derived from that filament (*see* polysiphonous)

perithallus: the upper, corticated region of a crustose plant; usually distinguished by plastid bearing cells

phycobilins: a group of protein-bound, water-soluble pigments found in members of the Cyanophyta and Rhodophyta, transmitting red and blue wavelengths

pinnate: with leaflets or filaments on opposite sides of an axis; featherlike

pinnules: ultimate fine branchlets arranged on a branch (see *Caulerpa*)

pit connection: a distinct wall connection between cells found in members of the Rhodophyta; junction may be a large (2-5 μm in diameter), cytoplasmic strand or a specialized cell wall plug between cells

plankton: small, free-floating organisms

plastids: cell organelles that function in photosynthesis

plurilocular: (term restricted to the Phaeophyta) applied to many-celled reproductive organs in which each cell produces a single reproductive body

pluriseriate: more than one cell broad (*see also* multiseriate)

pneumatophores: specialized erect roots of the black mangrove (*Avicennia germinans*) thought to function in gas exchange; these roots project through the mud or sand and provide an intertidal substrate for algal attachment

polysiphonous: with transverse tiers of vertically elongated cells (pericentral cells), usually arranged radially around a central, single-celled axis from which pericentral cells arise

proliferous: producing many branchlets

prokaryote: organisms that lack membraned organelles, e.g., bacteria, blue-green algae

propagule: specialized hook or spinelike branchlets, of determinate growth designed for asexual reproduction (see *Sphacelaria*)

prostrate: lying flat, procumbent morphology

protuberance: a projection

pyrenoid: a small, rounded, colorless body in a chloroplast functioning as a center of starch accumulation *pyriform:* pear shaped; kidney shaped

quadrate: in the form of a cross, cell walls formed at right angles

ramuli/ramulus: a final or ultimate branch or branches; term similar to branchlet in usage (see *Dasya*)

receptacle: swollen branches bearing the sexual or fruiting structures (conceptacles of *Sargassum*) found in upper portions of mature plants

rhizine: slender, thick-walled filaments

rhizoid: a colorless, one to many-celled filament serving as an organ of attachment

rhizome: a prostrate, creeping or underground stem (as in *Thalassia*); term also used to describe the creeping, horizontal portion of an alga (*Caulerpa*)

salinity: a term to describe salt concentration of seawater: usually measured by grams of salt per kilogram of water and expressed in parts per thousand (0/00)

scutate: lobed and flattened, applied to crustose, circular plant

secund: bearing appendages in a row along one side of a branch; not as regular as pectinate; sometimes the branches incurved

segment: a joint or a node (*see* articulated)

seirospores: a chain of thick-walled cells (*Seirospora*)

septate: transverse partitions present

septum: partition of cells across a lumen or a cross wall of a multinucleated cell

serrate: toothed like a saw and with the teeth pointing toward the apex

sessile: without a stalk; lacking a distinct supporting structure

seta: a hair or bristle, colorless, and projecting from the cell

sheath: a covering or coating in addition to the normal plant cell wall (Cyanophyta)

siphonous: (siphonaceous) tubular throughout and the filament but one cell, lacking cross walls; coenocytic

sorus: a group or cluster of reproductive organs, may be found in specialized structures (nemathecia or conceptacles)

spatulate: oblong and with the basal end narrowed; shaped like a spatula

spermatangium: the male sex organ producing the spermatia in the Rhodophyta

sporangia: a cell or cells within which spores are produced

spore: a specialized motile or nonmotile cell that becomes free from the parent plant and is capable of developing into a new plant

sporophyte: the diploid, spore-producing generation developed from a zygote

stellate: starlike

stipe: the upright, stemlike portion of a plant

sublittoral: region of ocean ranging from lowest tidal level to the edge of the continental shelf (or about 200 meters in depth)

supralittoral: region which is under direct influence of marine waters usually caused by splash zone of waves; above the high tide mark

terete: cylindrical; round in cross section

terminal: pertaining to the apex of axes or segments

tetrahedral: having contents of a tetrasporangium divided so that only three of the tetraspores can be seen in one view; the planes of division perpendicular (if segments triangularly divided, then pyramidal)

tetraspore: a spore formed within a tetrasporangium, usually one of four in a distinct division pattern (cruciate, zonate, et cetera)

tetrasporophyte: a diploid generation (in Rhodophyta) producing tetraspores.

thallus: term used to denote an undifferentiated plant body

tomentose: hairy, covered with filaments, spines, or hairs

trabeculae: slender strands of cell wall material extending into the cell of *Caulerpa*

trichoblast: a nonpigmented filament growing from the surface of a plant

trichocyte: a cell that develops into a hair cell

trichome: a row of cells connected end to end, a term used in blue-green algae to distinguish cells from the sheath

trichothallic: meristematic growth being confined to apex of plant with dividing cells producing a short filament and the plant body in alternate divisions apically and basally

trichotomous: hairlike, filamentous construction

truncate: squared off at the ends

ultimate branches: the final branching system (*see* branchlets; ramuli, ramulus)

uniaxial: with a single longitudinal filament through the center of a multicellular thallus (see *Hypnea*)

unilocular: term applied to one-celled sporangia producing many spores

uniseriate: a plant body or branch consisting of but one series of cells (filaments of *Cladophora*)

utricle: the swollen end of a filament forming a cortical stratum at the surface (see *Codium*)

vacuole: a space within a cell bounded by a membrane and filled with a watery substance

verticillate: arranged in whorls, or seemingly so

vesicle: a general term used to describe a bladder or bubble-shaped object or cell, (e.g., *Sargassum* floats, *Botryocladia* lateral branchlets).

whorled: surrounding a branch in a ring

xerophyte: a plant adapted to growth under dry conditions where water is unavailable or difficult to attain

zonate: having contents of a tetrasporangium divided by three parallel planes so that the tetraspores lie one above another; in other algae, thallus or markings on a thallus appear in distinct zones (see *Padina*)

zygote: a cell resulting from the union of two gametes; may be a persistent cell with thick, spiny wall

Bibliography

Arber, A. 1920. Water plants, a study of aquatic Angiosperms. Cambridge Univ. Press, London. 430 pp.

Ballantine, D. 1972. Algal epiphytes of four Florida seagrass species from Tarpon Springs, Florida. M.A. Thesis. Univ. South Florida, Tampa. 87 pp.

Barilotti, C. 1970. Non-genetic morphological variation in *Caulerpa prolifera*. M.A. Thesis. Univ. South Florida, Tampa. 62 pp.

Børgesen, F. 1913-1920. The marine algae of the Danish West Indies. I. Chlorophyceae. Dansk Bot. Arkiv. 1:1-158 (1913). II. Phaeophyceae. Dansk Bot. Arkiv. 2:1-66 (1914). III. Rhodophyceae. Dansk Bot. Arkiv. 3:1-435 (1915-1920).

Bowman, H. H. M. 1917. Ecology and physiology of the red mangrove. Proc. Amer. Philos. Soc. 56:589-672.

Bowman, H. H. M. 1923. The distribution and pollination of certain seagrasses. Mich. Acad. Sci. Arts Let. 2:3-10.

Chapman, V. J. 1954. The Siphonocladales. Bull. Torrey Bot. Club 81: 76-82.

Davis, J. H. 1940. Ecology and geologic role of mangroves in Florida. Carnegie Inst. Wash. Publ. No. 517; Pap. Tortugas Lab. 32:303-412.

Davis, J. H. 1943. The natural features of southern Florida, especially the vegetation of the Everglades. Geol. Bull. Fla. Geol. Survey 25:1-311.

Dawes, C. J. 1965. Algae and soil development in mangrove communities. Paper presented to Florida Academy of Science, April 1965.

Dawes, C. J. 1967. Marine algae in the vicinity of Tampa Bay, Florida. Univ. South Florida, Tampa. 105 pp.

Dawes, C. J. 1969. A study of the ultrastructure of a green alga, *Apjohnia laetevirens* Harvey with emphasis on cell wall structure. Phycologia 8:77-84.

Dawes, C. J., and J. F. van Breedveld. 1969. Benthic marine algae. Memoirs of the Hourglass Cruises. Vol. 1, Part 2, pp. 1-47. Florida Dept. Conservation Marine Laboratory.

Dawes, C. J., S. A. Earle, and F. C. Croley. 1967. The offshore benthic flora of the southwest coast of Florida. Bull. Mar. Sci. 17:211-231.

Dawes, C. J. and A. C. Mathieson. 1972. A new species of *Pseudocodium* (Chlorophyta, Siphonales) from the west coast of Florida. Phycologia 11:273-277.

Dawson, E. Y. 1966. Marine botany: an introduction. Holt, Rinehart & Winston, Inc. New York. 371 pp.

den Hartog, C. 1970. Seagrasses of the world. North-Holland Publishing Co., Amsterdam. 280 pp.

Desikachary, T. V. 1959. Cyanophyta. I.C.A.R. Monographs on Algae. Indian Council of Agricultural Research. New Delhi. 686 pp.

Dixon, P. S., and L. M. Irvine. 1970. Miscellaneous notes on algal taxonomy and nomenclature III. Bot. Notiser. 123: 474-487.

Dragovich, A., J. Finucane, and B. Z. May. 1961. Counts of red tide organisms, *Gymnodinium breve*, and associated oceanographic data from Florida west coast, 1957-1959. U.S. Fish Wildl. Serv. Spec. Sci. Rep. 369. Washington. 175 pp.

Dragovich, A., and J. A. Kelly, Jr. 1964. Ecological observations of macroinvertebrates in Tampa Bay, Florida. Bull. Mar. Sci. Gulf Carib. 14:74-102.

Dragovich, A., and B. Z. May. 1962. Hydrological characteristics of Tampa Bay tributaries. U.S. Fish Wildl. Serv. Fish. Bull. 62:163-176.

Drouet, F. 1968. Revision of the classification of the Oscillatoriaceae. Monograph 15. The Academy of Natural Sciences of Philadelphia, Philadelphia. 370 pp.

Drouet, F., and W. A. Dailey. 1956. Revision of the coccoid Myxophyceae. Butler Univ. Botan. Stud. 12:1-218.

Earle, S. A. 1969. Phaeophyta of the eastern Gulf of Mexico. Phycologia 7:71-254.

Egerod, L. E. 1952. An analysis of the siphonous Chlorophycophyta. Univ. Calif. Pub. Bot. 25:325-454.

Eiseman, N. J. (in preparation). Marine algae of Point of Rocks, Florida. Ph.D. Thesis, Univ. South Florida, Tampa.

Eubank, L. 1946. Hawaiian representatives of the genus *Caulerpa*. Univ. Calif. Publ. Bot. 18:409-432.

Fagerberg, W. 1972. A light and electron microscopic study of the gametophyte and sporophyte plants of *Padina vickersiae* Hoyt. M.A. Thesis, Univ. South Florida, Tampa. 50 pp.

Goodell, H. G., and D. S. Gorsline. 1961. A sedimentological study of Tampa Bay, Florida. Int. Geol. Congr. XXI Sess. Norden. 1960, Pt. XXIII:75-88.

Heald, E. J. 1969. The production of organic detritus in a south Florida estuary. Ph.D. Thesis. Univ. Miami, Miami, 110 pp.

Hine, A., and H. J. Humm. 1971. *Caulerpa ollivieri* in the Gulf of Mexico. Bull. Mar. Sci. 21:552-555.

Humm, H. J. 1956. Seagrasses of the northern Gulf Coast. Bull. Mar. Sci. Gulf Carib. 6:305-308.

Humm, H. J. 1963. Some new records and range extensions of Florida marine algae. Bull. Mar. Sci. Gulf Carib. 13:516-526.

Humm, H. J. 1964. Epiphytes of the seagrass, *Thalassia testudinum*, in Florida. Bull. Mar. Sci. Gulf Carib. 14:306-341.

Humm, H. J. Unpublished manuscript. The marine algae of Bimini. I. The Cyanophyta. 78 pp. 15 pls. (Manuscript in Univ. of South Florida Library, Tampa).

Humm, H. J. 1973a. III. The biological environment. A. Salt marshes, pp. IIIA 1-5. *In* A summary of knowledge of the eastern Gulf of Mexico. State University System of Florida Institute of Oceanography.

Humm, H. J. 1973b. III. The biological environment. B. Benthic algae of the eastern Gulf of Mexico, pp. IIIB 1-13. *In* A summary of knowledge of the eastern Gulf of Mexico. State University System of Florida Institute of Oceanography.

Humm, H. J. 1973c. III. The biological environment. C. Seagrasses, pp. IIIC 1-10. *In* A summary of knowledge of the eastern Gulf of Mexico. State University System of Florida Institute of Oceanography.

Humm, H. J. 1973d. III. The biological environment. D. Mangroves, pp. IIID 1-6. *In* A summary of knowledge of the eastern Gulf of Mexico. State University System of Florida Institute of Oceanography.

Humm, H. J., and S. E. Taylor. 1961. Marine Chlorophyta of the upper west coast of Florida. Bull. Mar. Sci. Gulf Carib. 11:321-380.

Jackson, C. R. 1952. Topographic and edaphic factors affecting plant zonation in a tidal marsh. Quart. J. Fla. Acad. Sci. 15:136-146.

Joly, A. B. 1954. The genus *Bostrychia* Montagne 1838 in southern Brazil, taxonomic and ecological data. Bol. Filosofia Cienc. Let. Univ. Sāo Paulo 173:53-74.

Jones, J. A. 1968. Primary productivity by the tropical marine turtle grass, *Thalassia testudinum* König, and its epiphytes. Ph.D. Thesis. Univ. Miami, Miami. 196 pp.

Kurz, H., and H. Wagner. 1954. Tidal marshes of the Gulf and Atlantic coasts of northern Florida and Charleston, South Carolina. Fla. State Univ. Studies 24:1-168.

Lewin, R. A., ed. 1962. Physiology and biochemistry of algae. Academic Press, New York. 929 pp.

Moore, D. R. 1963. Distribution of the seagrass *Thalassia* in the United States. Bull. Mar. Sci. Gulf Carib. 13:329-342.

Papenfuss, G. F. 1951. Notes on South African marine algae III. South African Bot. 17:167-188.

Phillips, R. C. 1960a. Ecology and distribution of marine algae found in Tampa Bay, Boca Ciega Bay and at Tarpon Springs, Florida. Quart. J. Fla. Acad. Sci. 23:222-260.

Phillips, R. C. 1960b. The ecology of marine plants at Crystal Bay, Florida. Quart. J. Fla. Acad. Sci. 23:328-337.

Phillips, R. C. 1962. Distribution of seagrasses in Tampa Bay, Florida. Fla. Bd. Conserv. Mar. Lab. Spec. Sci. Rpt. 6:1-12.

Phillips, R. C. 1967. On species of the seagrass *Halodule* in Florida. Bull. Mar. Sci. 17:672-676.

Phillips, R. C. and V. G. Springer. 1960. Observations on the offshore benthic flora in the Gulf of Mexico off Pinellas County, Florida. Amer. Midland Natur. 64:362-381, 4 tables.

Post, E. 1936. Systematische und pflanzengeographische Notizen zur *Bostrychia-Caloglossa Association.* Rev. Algol. 9:1-84.

Prescott, G. W. 1968. The algae: a review. Houghton Mifflin Co., New York. 436 pp.

Rehm, A. In preparation. Marine algal epiphytes of *Rhizophora mangle* prop roots along the west coast of Florida. Ph.D. Thesis. Univ. South Florida, Tampa.

Rydberg, P. A. 1909. The flowers and fruits of the turtle grass (*Thalassia*). J. N. Y. Bot. Gard. 10:261-265.

Sears, J. R., and R. T. Wilce. 1970. Reproduction and systematics of the marine alga *Derbesia* (Chlorophyceae) in New England. J. Phycol. 6:381-392.

Strawn, K. 1961. Factors influencing the zonation of submerged monocotyledons at Cedar Key, Florida. J. Wildl. Manage. 25:178-189.

Sykes, J. E. 1964. Requirements of Gulf and South Atlantic estuarine research. Proc. Gulf Carib. Fish. Instit., 16th annual session, pp. 113-120.

Tanner, W. F. 1960. Florida coastal classification. Trans. Gulf Coast Ass. Geol. Soc. 10:259-266.

Taylor, S. E. 1966. Phaeophyta of the eastern Gulf of Mexico. Ph.D. Thesis, Duke University, Durham. 312 pp.

Taylor, W. R. 1928. The marine algae of Florida with special reference to the Dry

Tortugas. Carnegie Inst. Wash. Publ. No. 379; Pap. Tortugas Lab. 28:1-219.

Taylor, W. R. 1936. Marine algae from the Yucatán Peninsula. Carnegie Inst. Wash. Publ. 461:115-124.

Taylor, W. R. 1954a. Distribution of marine algae in the Gulf of Mexico. Pub. Mich. Acad. Sci., Arts Letters 39:85-109.

Taylor, W. R. 1954b. Sketch of the character of the marine algal vegetation of the shores of the Gulf of Mexico, pp. 177-192. In P. S. Galstoff (ed.), Gulf of Mexico, its origin, waters and marine life. Fish. Bull. 89. Washington.

Taylor, W. R. 1957. Marine algae of the northeastern coast of North America. Univ. Mich. Press. Rev. ed. pp. 1-509.

Taylor, W. R. 1960. Marine algae of the eastern tropical and subtropical coasts of the Americas. Univ. Michigan Press, Ann Arbor. 870 pp.

Thomlinson, P. B., and G. A. Vargo. 1966. On the morphology and anatomy of turtle grass, *Thalassia testudinum*. I. Vegetative morphology. Bull. Mar. Sci. 16:748-761.

Tilden, J. 1910. Minnesota algae. I. The Myxophyceae of North America and adjacent regions including Central America, Greenland, Bermuda, the West Indies and Hawaii. Univ. Minnesota Bot. Ser. 8. 328 pp., 20 pls.

U.S. Coast and Geodetic Survey. 1960. Surface water temperature and salinity, Atlantic coast, North and South America. Pub. 31-1.

van den Hoek, C. 1969a. Notes on *Cladophora* (Chlorophyceae). I. Two new Caribbean species: *Cladophora jongiorum* nov. sp. and *Cladophora longicellulata* nov. sp. J. Phycol. 5:128-133.

van den Hoek, C. 1969b. Notes on *Cladophora* (Chlorophyceae). II *Cladophora catenata* (Linnaeus) Kutz., 1843, Emend V. D. Hoek, 1963. J. Phycol. 5:134-136.

Wynne, M. J., and W. R. Taylor. 1973. The status of *Agardhiella tenera* and *Agardhiella baileyi* (Rhodophyta, Gigartinales). Hydrobiologica 43:93-107.

Index